AN INTRODUCTION TO
MATHEMATICAL MODELS IN
ECOLOGY AND EVOLUTION

T0188710

ECOLOGICAL METHODS AND CONCEPTS SERIES

This series, successor to the Methods in Ecology series edited by John Lawton and Gene Likens, presents the latest ideas and techniques across the whole field of ecology and their application, from genetic to the global, from pest management to policy development. Books may be single- or multi-authored and will address emerging new areas within the field as well as updating well-established areas of endeavour. The new Series Editor is Professor Roger Kitching of Griffith University, Brisbane, who will welcome suggestions for works within the series. Email: r.kitching@griffith.edu.au

Ecological Methods and Concepts Series

Stable Isotopes in Ecology and Environmental Science
Second edition, 2007
Edited by Robert Michener and Kate Lajtha

An Introduction to Mathematical Models in Ecology and Evolution: Time and Space
Second edition, 2009
Michael Gillman

Forthcoming

Vegetation Classification and Survey
Andrew Gillison

Litter Decomposition in Aquatic Ecosystems
Edited by Mark Gessner

Canopy Science: Concepts and Methods
Edited by John Pike and James Morison

Methods in Ecology Series

Insect Sampling in Forest Ecosystems
2005
Edited by Simon Leather

Molecular Methods in Ecology
2000
Edited by Allan J Baker

Population Parameters: Estimation for Ecological Models
2000
Hamish McCallum

Biogenic Trace Gases: Measuring Emissions from Soils and Water
1995
Edited by PA Matson and RC Harriss

Geographical Population Analysis: Tools for the Analysis of Biodiversity
1994
Brian A Maurer

An Introduction to Mathematical Models in Ecology and Evolution

Time and Space

SECOND EDITION

MICHAEL GILLMAN

Department of Biological Sciences
The Open University
Walton Hall
Milton Keynes
MK7 6AA
UK

WILEY-BLACKWELL

A John Wiley & Sons, Ltd., Publication

Blackwell Publishing was acquired by John Wiley & Sons in February 2007. Blackwell's publishing program has been merged with Wiley's global Scientific, Technical and Medical business to form Wiley-Blackwell.

Registered office: John Wiley & Sons Ltd, The Atrium, Southern Gate, Chichester, West Sussex, PO19 8SQ, UK

Editorial offices: 9600 Garsington Road, Oxford, OX4 2DQ, UK
　　　　　　　　　The Atrium, Southern Gate, Chichester, West Sussex, PO19 8SQ, UK
　　　　　　　　　111 River Street, Hoboken, NJ 07030-5774, USA

For details of our global editorial offices, for customer services and for information about how to apply for permission to reuse the copyright material in this book please see our website at www.wiley.com/wiley-blackwell

Library of Congress Cataloguing-in-Publication Data
Gillman, Michael.
　An introduction to mathematical models in ecology and evolution : time and space / Michael Gillman. – 2nd ed.
　　　p. cm. – (Ecological methods and concepts series)
　Rev. ed. of: An introduction to ecological modelling / Michael Gillman, Rosemary Hails. 1997.
　Includes bibliographical references and index.
　ISBN 978-1-4051-7515-9 (pbk. : alk. paper) – ISBN 978-1-4051-9489-1 (hbk. : alk.
paper)　1. Ecology–Mathematical models.　2. Evolution (Biology)–Mathematical models.
I. Gillman, Michael. Introduction to ecological modelling.　II. Title.
　QH541.15.M3G5 2009
　577.01'5118–dc22

　　　　　　　　　　　　　　　　　　　　　　　　　　　　　　　　　2008046481

A catalogue record for this book is available from the British Library.

Set in 9.5 on 12 pt Meridien by SNP Best-set Typesetter Ltd., Hong Kong
Printed and bound in Malaysia by Vivar Printing Sdn Bhd

1　2009

Contents

Preface

This book is an introduction to the key methods and underlying concepts of mathematical models in ecology and evolution. It is intended to serve the needs of a broad range of undergraduate and postgraduate ecology and evolution students who need to access the mathematical and statistical modelling literature essential to their subjects. It assumes minimal mathematics and statistics knowledge (see below) while covering a wide variety of methods, many of which are at the forefront of ecological and evolutionary research. The book will also highlight the applications of modelling to practical problems such as sustainable harvesting and biological control.

There are many other ways in which this book could have been written and you will find examples of quite different treatments of modelling in the literature. In particular the book could focus on (and be lead by) applications, for example by asking whether models are helpful in understanding climate change or saving cod populations or reducing the incidence of malaria. The answer is yes to all of these but it was felt that it is better to try to understand the general principles underlying the models and then examine the applications. Doubtless my ideas of synthesis and generality are not those of others but it is an attempt to detect and reveal order. I also wanted to write a book with a lighter touch and so have avoided writing lengthy descriptions of method. Hopefully this makes the book accessible to a wider readership.

Understanding of the text will be helped by a familiarity with the basics of the following mathematical and statistical methods and concepts:

- manipulation of algebraic equations,
- logarithms and powers,
- differentiation,
- variance and standard error,
- significance and hypothesis testing.

Many thanks to Hils and Ed for encouragement and valuable comments. I have learnt much from interactions with colleagues at the Open University and previously at Imperial College. My interest in mathematical models was inspired by the lectures of Brian Goodwin and John Maynard Smith.

Michael Gillman
September 2008

CHAPTER 1

Introduction

1.1 What is a model?

A model is some representation of reality. In everyday life we are familiar with physical models of reality such as toy cars and film sets. Physical models have also been used extensively in science, perhaps most famously as the ball-and-stick model of DNA by Crick and Watson. Such physical models have been largely replaced by computer images and/or mathematical representations. In ecology and evolution the models are almost entirely of a mathematical nature. Reference to models and modelling in this book can therefore be read as *mathematical* models or modelling. Fortunately we do not have to be highly skilled mathematicians to construct and use such models. This text will show how to develop ecological and evolutionary models which have a wide application across the life sciences and are relevant to many other branches of science. Indeed, the understanding of modelling benefits from interplay between the sciences.

An important first step is to understand the overlap between statistical analysis and mathematical modelling. One of the central aims of statistical analysis is the description of trends and distributions in sets of data. For example, we might wish to provide a description of the change over time in the average size of a population. We can do this by using the statistical method of regression which provides a mathematical description of a line or curve of best fit through the data. The equation of the resulting line or curve is also a mathematical model of the population and could be used to predict change over time, with the possibility of extrapolating beyond the last time point. Of course, we need to be cautious over the extent to which extrapolations are performed, or, at the very least, make the user aware of the possible problems. Regression is introduced in the next section. A mathematical summary that allows prediction and extrapolation is therefore one valuable use of models. Expressing the model as a mathematical formula provides a brevity and formality of description. It also allows manipulation of the model and provides the opportunity for discovery of emergent properties not apparent from non-mathematical reasoning. Another related use of models is that they allow a simplification of reality. An alternative to producing models from observed temporal or spatial data is that we can build them from observed processes within the study system – for example, levels of survival or migration – and test them against observed data. Both of these modelling activities will be described in this book.

Ecological and evolutionary patterns in space and time are intrinsically mathematical. That is, events at one point in time or space can be related to events at previous points with the use of mathematical operators. This idea is illustrated by the Fibonacci sequence which starts with the numbers 1 and 1 and then continues by adding the previous two numbers to produce the sequence 1, 1, 2, 3, 5, 8 and so on. The sequence was named after Leonardo of Pisa who was known as Fibonacci (*c.* 1170–1250) and published in his *Liber Abaci* in 1202. This simple arithmetic sequence was originally used to predict changes in rabbit numbers and so is an early, perhaps the earliest, example of an ecological model. Although the underlying assumptions about how rabbits reproduce were naïve – for example, that female rabbits always produce one pair of rabbits every month, which contains one male and one female – they did consider appropriate biological issues of sex ratio and delayed reproduction (female rabbits were assumed to start reproducing after 2 months). In the model the first two numbers of the sequence (1, 1) represent one pair of rabbits in months 1 and 2. At the start of month 3 one new pair is born so the total is 2. At the start of month 4 another pair is born to the first pair making the total 3 and at the start of month 5 the original pair again reproduces as does the second pair (total of 5). The model sequence generates an increase with time which, at least in the short term, looks something like population increase (Fig. 1.1; we will see similar examples of population increase later).

This very simple model of rabbit population dynamics can be contrasted with modern complex models which incorporate features such as disease

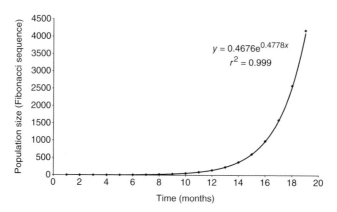

Fig. 1.1 Increase in population size represented by the Fibonacci sequence increase. The fit of the curve through the Fibonacci sequence is discussed in Section 1.4.

(myxomatosis, rabbit haemorrhagic disease), variation in numbers of off-spring per litter and random variation (Scanlan et al. 2006).

In the next two sections we will introduce the statistical methods used to generate mathematical models from observed data sets; that is, the first of the two types of modelling approach described above.

1.2 Regression and its use in mathematical modelling

Regression techniques are used widely in science for determining the significance and shape of relationships between two or more variables. An important distinction is between correlation and regression. While correlation measures the strength of a relationship between variables, regression also aims to describe the relationship between two or more variables. The use of regression implies that there is a causal relationship between the variables; that is, that there is an independent and dependent variable. Changes in the independent variable may be generated in an experiment, for example by varying temperature and measuring the response in a dependent variable such as chemical reaction time, or naturally generated, for example due to different climatic conditions. In regression analysis with two variables, the dependent variable is plotted on the vertical axis and the independent variable on the horizontal axis. Regression fits a line through the data (a line of best fit), which is a mathematical representation of the relationship between the variables (Fig. 1.2). If the causal relationship is linear then the equation of the line is $y = mx + c$, where y and x are the dependent and independent

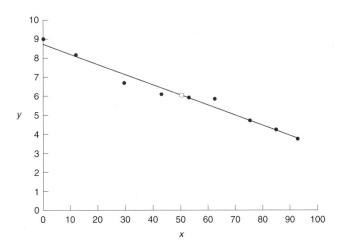

Fig. 1.2 A line of best fit through a set of data points (x, y) will pass through the means of the x and y values (indicated by the square symbol in the centre).

variables respectively and m and c are the gradient and y intercept. This is a mathematical model of the data set, allowing a prediction of y for a given value of x. An important feature of regression analysis is that it allows an estimate of the mean values of the parameters m and c and the reliability of those estimates, indicated by the standard error (SE) of the estimate. The same principle applies to more complex regressions.

Significance testing can be used in various ways in regression. The most basic question is whether the dependent variable changes with the independent variable. The null hypothesis of no change is represented by a horizontal line; that is, a gradient of zero. Therefore, in a linear model, a test of whether the gradient is different from zero is also a test of whether the overall regression model is significant. Regression analysis uses the same principles for significance testing as analysis of variance (ANOVA). In regression the sum of squares (SS) is determined from the sum of the squared differences between an observed dependent value and its corresponding predicted value on the regression line at the same independent value. It is helpful to imagine the regression line pivoting through the mean of the dependent and independent values (Fig. 1.2). When the line is horizontal, the SS of the observed values is equal to the total SS because all the predicted values are the same as the mean of the dependent value. As the line pivots, the SS reduces until the line of best fit is reached. The difference (or deviation) of an observed value from the predicted value in the regression is called the residual. From the residuals we calculate an error or residual SS, so called because it is not explained by the regression. The difference between the total SS and the error SS is the regression SS; that is, the amount explained by the regression. Thus the SS terms are additive:

Total SS = regression SS + error SS

The regression SS and error SS are divided by their respective degrees of freedom to give the regression and error mean squares (MS). The significance of the regression is then measured by the relative sizes of the regression MS and the error MS (F = regression MS/error MS), just as with the F test in ANOVA. An example is given below.

Significance testing also applies to the parameters of the regression model. In the linear example the overall significance is the same as the significance of the parameter m because we are usually looking for significant departures from the horizontal; that is, a value of zero for m. However, it is possible that our null hypothesis requires tests of departures from gradients other than zero. We will see later how significant parameter values can be used in a variety of ecological and evolutionary models.

The method of linear regression will be illustrated with a grassland ecosystem example. A study by Hui and Jackson (2006) explored the relationship between climate variables and above- and below-ground biomass in different grassland ecosystems across the world. The examples came from

Asia, North and Central America and Africa. The authors' main interest was in the fraction of the total net primary production (NPP) below ground and its contribution to global carbon cycling. Here we will use their data on above- and below-ground NPP and mean temperature and rainfall at the different sites (Fig. 1.3). Note that the units of NPP are g of dry matter (DM) m^{-2} that accumulated over a fixed period of time; that is, it is a measure of the growth of the grassland species above and below ground.

Several trends are suggested from the best-fit lines in Fig. 1.3. Below-ground NPP appears to decline with increasing temperature and rainfall

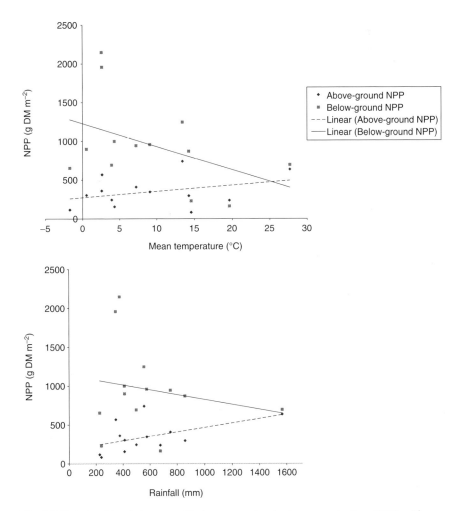

Fig. 1.3 Relationship of above- and below-ground net primary production (NPP) with temperature and rainfall. The lines of best fit are determined by linear regression. DM, dry matter.

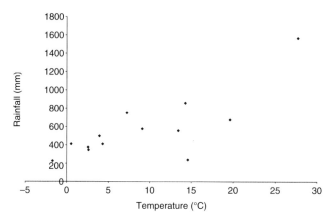

Fig. 1.4 Relationship between rainfall and temperature.

whereas above-ground NPP appears to increase with temperature and rain-fall. The distribution of points in the two graphs looks rather similar, which makes us suspect that temperature and rainfall are closely correlated (Fig. 1.4)

In fact the correlation coefficient is 0.777, which is statistically significant ($P < 0.01$). This high and significant correlation means that either rainfall or temperature can be used for the analysis. Notice that we have not fitted a line through the points in Fig. 1.4. This is because we do not wish to imply that one variable is dependent on the other.

Although trends are suggested in Fig. 1.3 we need to run the regression analysis to check the significance. The null hypothesis is that there is no change in the NPP with either rainfall or temperature. The results for above-ground NPP and temperature are $F = 1.435$, $P = 0.256$. Therefore the slope is not significantly different from zero. The regression analysis also provides an estimate of the values of m and c (gradient and intercept; the gradient is sometimes indicated by the letter b). However, this estimation is less useful when the regression is non-significant. In this case the value of m is 8.014 and the intercept is 272.39. Therefore, the regression equation would be:

Above-ground NPP = $(8.014 \times \text{temperature}) + 272.39$ (1.1)

Equation 1.1 could be used for predicting the value of above-ground NPP for a given value of temperature, but we would be wary of using it because the regression is not significant.

In fact, none of the four regressions in Fig. 1.3 are significant! This seems odd as we might expect temperature and/or rainfall to have a major effect on biomass. It may of course be that the non-significance is the correct result and that grassland NPP accumulation, when measured across different geo-graphical areas, is independent of climate. An inspection of the data suggests

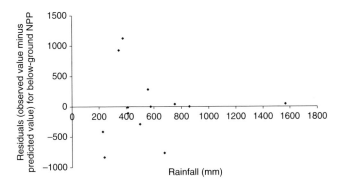

Fig. 1.5 Residuals of below-ground NPP from regression against rainfall.

that some of the problems may lie in the large amount of variation in the data around the regression. We can consider this by examining the residuals; that is, the size of the difference of the observed values from the predicted value on the regression line (Fig. 1.5)

The pattern of residuals of below-ground NPP with respect to the rainfall shows a large scatter of values (positive and negative) between approximately 200 and 700 mm of rainfall, with small residuals for the higher values thereafter. This is a common distribution of residuals and a source of concern as regression analyses require that the residuals follow the same distribution pattern across the full range of independent values. A consequence of this deviant pattern is that the larger values of rainfall have a large effect on the regression. Indeed, it is possible to have a completely non-significant scatter of points at low values and for one point with a high value (for both variables) to generate a significant positive regression. Generally this means that more samples are needed in the intermediate range. In this case there is one large value with a mean temperature above 25°C and rainfall of nearly 1600 mm (you can see this clearly on the plot of rainfall against temperature, Fig. 1.4). Not surprisingly this occurs in a tropical region (Thailand). If we remove this value the regressions are even further away from significance.

Whereas the NPP above or below ground did not show a significant response to the climate variables, the fraction of below-ground NPP was significant (Fig. 1.6).

This study illustrates the importance of exploring your data. Although above-ground and below-ground NPP show no significant effects, the fraction of NPP below ground declines significantly with increasing temperature. Alternatively, we could also use the fraction of NPP above ground (what slope would you expect for this regression?). The same effect is seen with rainfall, which is expected as rainfall and temperature are positively correlated. The relationship of fraction of below-ground NPP with temperature is given by the equation:

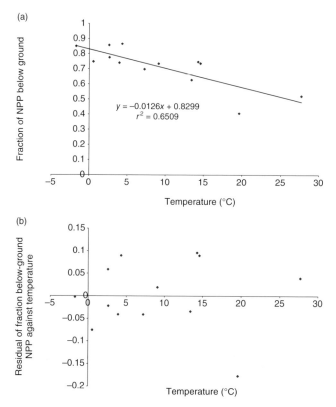

Fig. 1.6 (a) Relationship of fraction of below-ground NPP with temperature. We can also see that the distribution of the residuals with respect to temperature (b) is much better than in Fig. 1.5.

$$\text{Fraction of below-ground NPP} = (-0.01255 \times \text{temperature}) + 0.8299 \qquad (1.2)$$

Therefore for each increase of 1°C of mean temperature there is predicted to be a corresponding decrease of 0.01225 in the fraction of below-ground NPP. The high significance of the regression ($P < 0.001$) gives us greater confidence in using equation 1.2 over equation 1.1. Furthermore the residuals appear to be well behaved along the full range of independent values. However, there still appears to be a reasonable amount of scatter around the predicted line. The amount of variation explained by the regression is summarized by the value of r^2, the correlation coefficient squared. It has a particular meaning in regression analysis where it is known as the coefficient of determination. The r^2 indicates the fraction of variation in the dependent variable explained by variation in the independent variable. Therefore, in this example, an r^2 of 0.651 means that 0.651 (or 65.1%) of the variation in fraction of below-ground NPP is explained by variation in temperature. From a modelling perspective this may be sufficient to provide a useful model of the effects of climate change on grassland ecosystems. Conversely, about one-third of the variation remains unexplained. We will see in the next section how to improve these models by using more than one explanatory variable.

1.3 Multiple regression

In the previous section we had two possible independent variables which were correlated. This meant that either could be used and, indeed, it was only appropriate to use one. However, it is often the case that there are two or more independent variables that are not correlated and we wish to understand their overall contribution to the regression. This is the domain of multiple regression. With two independent variables, instead of fitting a line through a set of points as in linear regression, we are now fitting a two-dimensional plane through a set of points. With three independent variables it becomes difficult to visualize the process, but the mathematical principle of reducing the unexplained variation still applies.

We will now consider the application of multiple regression to an ecosystem analysis. The hypothesis was that methane production from wetlands was being suppressed by acid rain. As methane is a powerful greenhouse gas the reduction may have important consequences for global climate change. An experiment was undertaken involving treatments simulating acid-rain deposition on a peat bog with the levels of methane flux recorded (Gauci et al. 2002). A further analysis considered the relationship of the extent of methane flux with peat temperature and water-table depth; that is, two naturally occurring variables. This required use of multiple regression and is considered here.

The dependent variable is methane flux and the independent variables are peat temperature and peat water table. The data can be plotted as two separate graphs (Fig. 1.7) or combined in a three-dimensional graph in which the two independent axes are at right angles to each other (these may look impressive but it is often very difficult to read the values and so they are not included here).

The plots of the dependent against the independent values separately show that the percentage of methane flux increases with temperature and with water-table depth. Notice that the r^2 values of these two are quite low (30 and 24% of variance explained respectively) although the regressions are significant ($P = 0.0045$ and 0.012). There is also the suggestion that there is a problem with the distribution of the residuals (Fig. 1.8). There are ways of dealing with these issues without further sampling, such as transformations of data, which were undertaken by the authors of the paper: we will consider this later. Here we will focus on the raw data and ignore the odd patterns of the residuals.

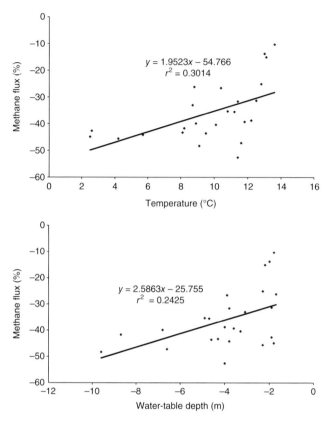

Fig. 1.7 Percentage of methane flux plotted against the two independent variables (peat temperature and water-table depth).

First, we check that there is no relationship between the (assumed) independent variables. It is true in this case although we note that there is a curious shape to the data with little variation in the water table at high and low temperatures but large variation at intermediate values (Fig. 1.9). Whereas it is possible to analyse the relationship between each independent variable and the dependent variable separately, we suspect that it is a combination of factors that contribute to the methane flux and we wish to capture that information in one mathematical statement. Rather than $y = mx + c$ we now have two independent variables so the overall equation will look like this:

$$\text{Methane flux} = (a \times \text{temperature}) + (b \times \text{water-table depth}) + c \qquad (1.3)$$

Or in general terms:

$$y = ax_1 + bx_2 + c \qquad (1.4)$$

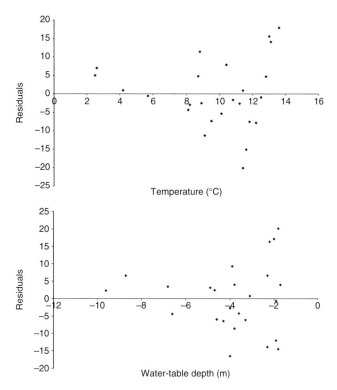

Fig. 1.8 Residuals from the linear regressions in Fig. 1.7.

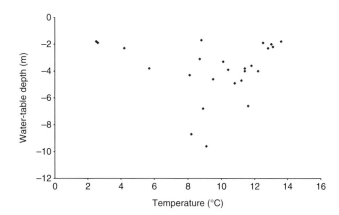

Fig. 1.9 Relationship between temperature and water-table depth. The correlation is not significant ($r = 0.02$, $P > 0.9$).

Like the simple regression counterparts these equations allow prediction of the dependent variable with given values of independent variables. Equation 1.3 indicates that for given values of temperature and water-table depth a value for methane flux can be predicted. The estimates of a, b and c are generated from the multiple regression analysis. Methane flux can then be modelled under a range of scenarios of temperature and water-table depth.

The multiple regression significance results are given in Table 1.1a with the parameter estimates (regression coefficients) detailed in Table 1.1b. The estimates are 1.99 and 2.65 for temperature (a in equation 1.3) and water-table depth (b in equation 1.3), respectively, in agreement with the predictions from the simple regressions. Notice that all the estimates are highly significant; that is, they are all highly significantly different from 0. This is not always the case in multiple regression. If variables are not significant then they are dropped from the analysis until we are left with the model (regression) that only contains significant components. The r^2 is also improved in this model (0.556) over the linear regression. This is expected as we have combined two significant elements which are independent. Thus 55.6% of variation in methane flux is predicted by variation in temperature and water table. The full predictive equation is therefore:

$$\text{Methane flux} = -44.8 + (1.99 \times \text{temperature}) + (2.65 \times \text{water-table depth})$$

In summary, multiple regression can be used as a tool for reducing complex models to their statistically significant components and for exploring the interplay between different explanatory variables. In this example we considered two linear relationships. The next section will address nonlinear

Table 1.1 (a) Regression model statistics and (b) parameter estimates and significance in methane flux regression model ($n = 25$).

(a)

	Sum of squares (SS)	Degrees of freedom (df)	Mean squares (MS; calculated as SS/df)	F (regression MS/residual MS)	Probability (P)
Regression	1656.311	2	828.1554	13.799	0.000131
Residual	1320.312	22	60.0142		
Total	2976.623				

(b)

	Regression coefficient	SE	t-test value	P
Intercept	−44.802	5.846	−7.664	<0.00001
Temperature	1.993	0.505	3.946	0.00069
Water-table depth	2.653	0.746	3.556	0.0018

regression methods and from there move into an important class of mathematical models which has been used to describe a wide range of scientific phenomena.

1.4 Nonlinear regression and power laws

To illustrate nonlinear regression let us return to the familiar example of the Fibonacci sequence. Imagine that the Fibonacci sequence is used to model change in population size over time. According to this model the population size at any point in time can be predicted given the starting values and the simple mathematical operation of adding the two previous values. In Fig. 1.1 we saw the increase in size over 19 time points. The striking result is that the values appear to increase slowly at first but then increasingly rapidly. This is typical of geometric or exponential increase in which the values at one time point are some multiple of the values at the previous time point. Geometric sequences are in contrast to arithmetic sequences where a term in the sequence is produced by adding a constant value to the previous term. In the Fibonacci sequence, the values at the current time are the sum of the values at the previous two time points, which simulates geometric growth. Geometric growth or geometric sequences in general can be summarized using the mathematical notation of powers. For example, the geometric sequence 1, 2, 4, 8, 16, 32 . . . is produced by starting with 1 and multiplying it by 2; this multiplication is then repeated. If the terms of the sequence are themselves numbered 0, 1, 2, 3, 4 . . . then the geometric sequence is seen as 2^0, 2^1, 2^2, etc. Furthermore, if the sequence term is represented by the letter t, then the sequence value for term t is simply 2^t. Population and evolution models of this form will be considered in Chapter 2.

In Fig. 1.1 a curve is fitted through the points with the following equation:

$$N_t = 0.4676e^{0.4778t} \tag{1.5}$$

where t is time and N_t is the population size at time t. This is an example of a nonlinear regression in which a curve is fitted through the points. The best fit is determined in a similar manner to a linear fit, for example by minimizing the squared differences between the observed and predicted values which we encountered with linear regression, a technique known as least squares. Comparison of equation 1.5 with 2^t suggests that it is a mathematical summary of a geometric sequence. Equation 1.5 is slightly more complex in that it uses the number e instead of 2, t is multiplied by the value 0.4778 and $e^{0.4778t}$ is multiplied by 0.4676.

The number e is a special (natural) number with particular properties and an illustrious history involving Napier, Mercator, Bernoulli, Leibniz and finally Euler, who gave a full description of the value of e to 18 decimal places

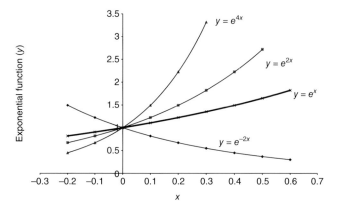

Fig. 1.10 Examples of exponential functions. Notice the effect of a negative power.

in 1748 (www-history.mcs.st-andrews.ac.uk/HistTopics/e.html). For example, if we plot the curve $y = e^x$ we discover that the gradient of the curve at any point x is equal to y (Fig. 1.10). Mathematically this is stated as $dy/dx = e^x$ (this type of equation is explained in detail in later chapters). e is an irrational number (like π) with a value of 2.718. . . .

Instead of fitting a curve through the points we can employ a mathematical transformation to convert an exponential curve to a straight line and then use linear regression. The appropriate transformation for exponential functions involves logarithms (abbreviated to logs). If $y = 10^x$ then taking the logarithm to the base 10 (written as $\log_{10}$) of both sides gives:

$$\log_{10} y = x$$

Similarly, if $y = 10^{ax}$ then $\log_{10} y = ax$. If the exponential function is of the form $y = b10^{ax}$ then taking logs gives $\log y = \log b + ax$. Therefore we can see that taking logs transforms a nonlinear function to a linear function. Plotting $\log y$ against x will give a straight line with $\log b$ as the intercept and a gradient of a.

Logarithms can be given to any base number. The base number e provides an important type of logarithm called natural or Naperian logarithms. Instead of writing $\log_e$ we use the abbreviation ln. So if we take the natural log (ln) of both sides of equation 1.5 we can produce a linear equation (i.e. one with an equation $y = mx + c$; Fig. 1.11):

$$\ln y = \ln (0.4676) + 0.4778x \tag{1.6}$$

This is a useful result as it allows us to use linear regression to estimate the parameters of the exponential equation. The straight line equation is $y = -0.7601 + 0.4778x$. We can see that 0.4778 agrees with the exponent or power value in equation 1.5; $\ln(0.4676)$ is also equal to -0.7601. The fit of the regression is very high, the r^2 has a value of 0.999 indicating that 0.999

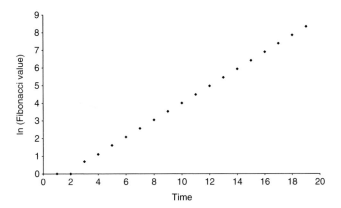

Fig. 1.11 Logarithmic transformation of a Fibonacci population model to produce a linear relationship.

or 99.9% of the variation in the dependent variable is explained by variation in the independent variable. This fit can be improved further by seeing that the first two values are high residuals (1 and 1 were fixed start points). Removing those two values and fitting through the remainder gives $y = -0.7938 + 0.4804x$ with an r^2 of 1. Actually, it is not quite a perfect fit ($r^2 = 0.9999$)!

Let us recap what we have done here. We have made a model of population dynamics with the Fibonacci series. This model can be described mathematically by either a nonlinear function ($y = ae^{bx}$) or an equivalent linear equation ($\ln y = \ln(a) + bx$). This means that we could predict the numbers of rabbits at any time. In doing this we have introduced several important methods: linear and nonlinear regression, exponential or geometric growth and logarithmic transformation. All of these methods will be used and explained in further detail later in the book.

Just as e was identified as a natural number it is clear that many biological and other scientific phenomena are naturally nonlinear. Let us consider a classic ecological example of a nonlinear relationship and then step back to consider why such relationships might arise. It has long been known that (other things being equal) habitats with larger areas contain more species. These areas may have natural boundaries, such as islands within an ocean, or be different sample areas within a larger area of suitable habitat. Whereas the increase of species number with habitat area is perhaps intuitive, the precise form of the relationship and interpretations of the mechanisms are not so straightforward (Fig. 1.12).

In general it has been shown that species/area relationships can be modelled as a power function:

$$S = cA^z \tag{1.7}$$

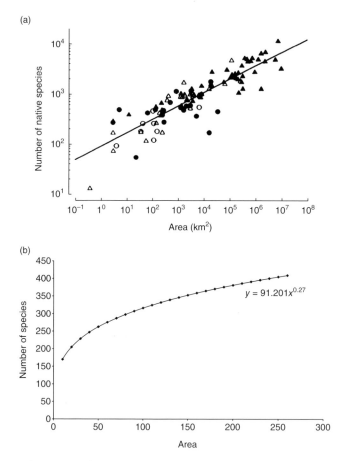

Fig. 1.12 Mathematical relationship of species on islands with island area. The data (a) are shown on a log-transformed linear plot. The shape of the curve without the logarithmic transformation is shown in (b). Note that the graph in (b) does not cover the full range of values in graph (a). Graph from Lonsdale (1999) reprinted in Williamson et al. (2001).

where A is area (independent) and S is the number of species (dependent). Note the difference between power functions as exemplified by equation 1.7 and the exponential function such as equation 1.5. In the case of the exponential function the independent variable (x) is the power, e.g. $y = a10^{bx}$ and the linear transformation is achieved by plotting $\log y$ against x. In the case of the power function the independent variable is the base (e.g. $y = ax^{b}$) with the corresponding linear transformation of $\log y$ against $\log x$. So in the case of $S = cA^{z}$ taking the logarithms of both sides yields:

$$\log S = \log c + z \log A \tag{1.8}$$

With this transformation we could regress $\log S$ against $\log A$ and determine z and $\log c$ from the gradient and intercept respectively. In Fig. 1.12a, the axes are log-transformed (the original untransformed values are given on logarithmic axes; equally one could show the log values, e.g. −1 to 8 on the area axis) showing a linear relationship between numbers of native species and island area. The fitted line is:

$$\log S = 1.96 + 0.27 (\log A) \tag{1.9}$$

Thus $z = 0.27$ and $\log c = 1.96$ (therefore $c = 10^{1.96} = 91.2$).

Species/area relationships fitted to power functions have been shown to occur across a variety of plant and animal groups leading to the suggestion that this is a natural ecological 'law'. Describing it as a law as opposed to an empirical generalization depends on whether there is a consistent underlying mechanism. The semantics of these debates are beyond the scope of this book but it is certainly the case that power functions of the general form $y = cx^b$ not only describe species/area relationships but also arise frequently across a range of scientific phenomena. These include various relationships with body mass including respiration (metabolic) rate, population density and generation time; variance/mean relationships applied to populations and numbers of cells in bodies and evolutionary processes such as species interactions through time.

Why should power functions be so prevalent? One answer is that power functions can occur when two variables are linked by a third common variable. To illustrate this answer we will consider a simple physical example.

Imagine a set of spheres of different sizes. The size of the sphere is given by the radius (r). We can calculate the surface area and the volume of each sphere (surface area $= 4\pi r^2$ and volume $= (4/3)\pi r^3$). If we then plot the surface area against the volume for each sphere we obtain a power function (Fig. 1.13a), which when log-transformed gives a linear function (Fig. 1.13b).

The equation of the power function is:

Surface area $= 4.836 \times \text{volume}^{2/3}$

which when log-transformed is:

Log surface area $= \log(4.836) + 2/3 \log \text{volume}$

The volume and surface area of a sphere are related through a common third variable (radius). We can see this from the ratio of surface area to volume:

Surface area of sphere/volume of sphere $= 4\pi r^2 / (4/3)\pi r^3 = 3/r$

As surface area is proportional to r^2 and volume is proportional to r^3, increasing the radius r will produce a power function which increases by r^2/r^3; that is, 2/3. So the value of the power is 2/3. If volume is plotted against surface area the power value would be 3/2. This reasoning can be applied to biological phenomena. Imagine that we are plotting the rate of heat loss of different

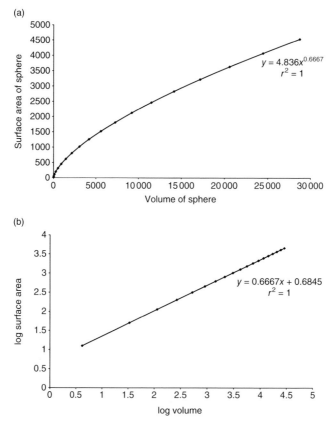

Fig. 1.13 Relationship of surface area and volume of spheres of different radius. (a) Power function and (b) corresponding log-transformed plots.

organisms against their mass. We might expect that heat loss will be proportional to the surface area of the organism. As mass will be proportional to volume, a power function with an exponent of about 2/3 would be expected. As organisms are generally not perfect spheres and have various physiological contraptions that, for example, reduce heat loss, deviations from 2/3 may occur. Similarly there will be errors in measurement which will contribute to variation around the regression line. However, you can see that power laws may be expected simply as a consequence of scaling of body size. These phenomena have been widely studied and are referred to as allometric processes.

The principle of a common third variable generating power functions through scaling can be applied to other ecological and evolutionary processes. Take the −3/2 thinning rule (or law) as an example. The −3/2 power law has been much discussed among plant ecologists as it was considered as one of the

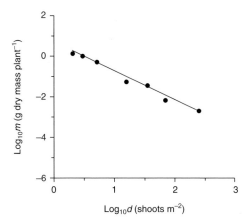

Fig. 1.14 Example of self-thinning rule with a marine alga (*Phyllariopsis purpurascens*). A significant frond mean mass (*m*) to density (*d*) relationship was obtained ($\log_{10} m = 0.6 - 1.4 \log_{10} d$) where the slope of −1.4 was not significantly different from the expected −1.5 (Flores-Moya et al. 1996).

few 'laws' in ecology. The phenomenon is observed when plants of a particular species are grown or occur naturally at different densities (Fig. 1.14). If the mean masses of plants are plotted against plant density then, as the plants grow, they reach and then follow a boundary which has a gradient of approximately −3/2. The following of the boundary (and therefore reduced density) indicates that plants are lost and therefore thinning.

An explanation of this rule is that mean plant mass is proportional to volume (l^3) whereas density is inversely proportional to area (l^2). Therefore density (N) = k/l^2 and mean plant mass (m) = pl^3 where k and p are constants. These two equations can be rearranged to make equations in terms of l:

$$l = (k/N)^{1/2}$$

$$l = (m/p)^{1/3}$$

This shows that:

$$(k/N)^{1/2} = (m/p)^{1/3} \qquad (1.10)$$

A power of 1/2 means a square root of a number whilst a power of 1/3 means the cube root. If we cube both sides of equation 1.10 we obtain:

$$\left((k/N)^{1/2}\right)^3 = m/p$$

$$\left(k^{1/2} N^{-1/2}\right)^3 = m/p$$

$$(k^{3/2})(N^{-3/2}) = m/p$$

$$m = p(k^{3/2})(N^{-3/2})$$

Combine p and $k^{3/2}$ into a new constant, s:

$$m = sN^{-3/2}$$

So, mean plant mass (m) is related to density (N) to the power $-3/2$. In fact, the $-3/2$ rule has been subsumed within a more general $-4/3$ rule relating average mass to maximum density (Enquist et al. 1998).

1.5 Conclusion

As you may be starting to appreciate, there are few areas of science which cannot benefit from an understanding of mathematical models or cannot be couched in mathematical terms. The introductory methods covered in this book should allow you to start developing your own models and provide an insight into the methods commonly used by a wide variety of practitioners. Hopefully you will begin to appreciate the wide range of applications, including biological control, sustainable harvesting and conservation management.

An important distinction in mathematical modelling, and one which informs the organization of the next two chapters, is between deterministic and stochastic processes. In a deterministic world everything should be predictable. For example, if population dynamics are deterministic we should be able to predict the population size at time t given a knowledge of the processes (described by mathematical equations) underlying the dynamics. Simple models with deterministic dynamics are the subject of Chapter 2. This notion of the deterministic world is undermined in two distinct ways. First, deterministic processes do not necessarily lead to predictable outcomes (as we shall see in Chapter 5) and second that stochastic or random events may be as important in ecological and evolutionary dynamics, as we will discover in Chapter 3. In most cases a combination of stochastic and deterministic modelling is the best way to proceed. To use the regression analogy, we need to identify and quantify the deterministic signal (the variation due to the regression) and we need to find ways of modelling the unexplained variance, which may be the result of extrinsic random events and/or sampling error. We also need to be aware that processes which are stochastic at one temporal or spatial scale may be much more predictable at larger scales. So, while it is difficult to predict the occurrence of storms from day to day, we may be much more certain about their occurrence and even their strength from month to month.

CHAPTER 2

Simple models of temporal change

2.1 Introduction

In this chapter we will begin constructing some simple models which describe change in an ecological or evolutionary variable such as population size or number of species over time (change over time is referred to as *temporal* change). Similar models can be constructed to describe spatial change and this is the subject of Chapter 8. The models that we generate can be used to predict particular sizes of the study variables or particular properties of those variables. These properties include their tendency to return to certain values, the possibility of cycles and the movement to either very high or very low values, including zero (Fig. 2.1).

Although population outbreaks or species extinctions may grab the news it is also true that populations of a wide range of species can persist at similar levels for many years, whereas the fossil record demonstrates the persistence of groups of similar species over many millions of years. Why do some populations fluctuate enormously while others persist at or around a particular size over time? What governs the dynamics of populations? Why do some species persist while others go extinct? Clearly, modelling cannot answer all of these questions, but it can help in identifying some of the major processes and help quantify the different levels of fluctuation and the likelihood of certain dynamics including the end point of extinction. This chapter starts off the investigation into stability in temporal models that concludes in Chapter 5.

Modelling of temporal dynamics can be used in various applications. For example, we may be interested in the long-term fluctuations in a herbivore population which is believed to a keystone species in an ecosystem. We may wish to harvest that herbivore in a sustainable manner; that is, without causing the herbivore to become locally extinct and thereby undermine the ecosystem. By modelling the changes in herbivore numbers over time we may be able to determine the levels of sustainable harvesting and predict whether the ecosystem of which it is part will persist. Similarly we may wish to determine how a pathogen such as the malaria parasite (various *Plasmodium* species) responds to the introduction of a control measure such as spraying of the mosquito vector (*Anopheles* species) or reduction in bite rate due to use of nets.

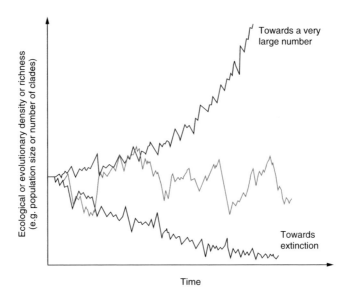

Fig. 2.1 Illustration of dynamics of ecological and evolutionary variables with time.

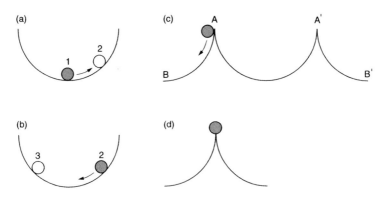

Fig. 2.2 Stability and equilibrium illustrated by a ball in a cup. (a) Displacement of ball from (apparent) equilibrium at position 1 to position 2. (b) Release of ball from position 2 (or equivalent position 3). (c) Displacement of ball beyond local stability boundary at A or A'. (d) Unstable equilibrium.

In order to pursue these lines of enquiry we need to understand some key terms: stability, equilibrium and perturbation. A simple physical model will illustrate these terms. Imagine that a ball is placed in a centre of a cup (Fig. 2.2). The ball is at rest but is it stable? We can only know this if we move the ball; that is, we perturb it. Upon release the ball returns to the base of the cup. Therefore we can say that the ball at the bottom of the cup is at a position of stable equilibrium, defined as the steady state to which the ball will return after perturbation. Stability is related to equilibrium in that it describes the tendency of a population or other system to stay at or move towards or around the equilibrium. However, the stability of this equilibrium depends on the degree of perturbation. If we push the ball beyond the edge of the cup it falls out of the cup and away from the stable local equilibrium (Fig. 2.2c). The equilibrium is therefore locally stable but not globally stable. The same ideas can be applied to ecosystems or components of ecosystems, such as populations of herbivores or decomposers. Thus for a population or ecosystem the equilibrium or steady state can be defined as the state (e.g. density) to which the population or ecosystem returns after perturbation. The stability of the whole ecosystem can be considered with respect to energy flow, nutrient cycling or the interactions between its components. The distinction between local and global properties of stability is also important here. A population may persist under small amounts of perturbation from its equilibrium value but move towards extinction or outbreak conditions under larger perturbations.

There are other properties of the locally stable equilibrium that we might wish to consider, for example, the rate of return of the ball to the equilibrium after perturbation. The local stability in the physical model of Fig. 2.2 also relies on friction slowing the ball down after release (from position 2); otherwise, with the aid of gravity, it would be like a frictionless pendulum switching continuously from position 2 to 3.

Imagine a second physical model in which we balance the ball on the tip of a pin-head (Fig. 2.2d). The ball is at rest but the equilibrium is highly unstable – any very minor perturbation will send the ball off to another place. Such an unrealistic view of stability features in some ecological models, as you will appreciate later.

As with the physical model, we can reveal the stability boundaries of the ecological system by perturbation experiments. For example, consider two interacting species such as two competitors or a predator and prey species. Experimental perturbations of the densities of A or B may reveal any stability boundaries (Fig. 2.3). In the example of Fig. 2.3, reduction of the density of species A to value y results in return to the initial (equilibrium) density (x), whereas reduction to z pushes the system beyond the local stability boundary. This could be undertaken in the field as a removal experiment in which the density of species A is reduced by different amounts and its and other (possibly competing) species return to the apparent equilibrium investigated. There are many examples of such experiments involving both removal and addition of species. These studies have been complemented by investigations of the effects of altering abiotic components such as nitrogen levels. Often the results are viewed in terms of the whole community or ecosystem and so a fuller discussion of these studies is reserved for later chapters.

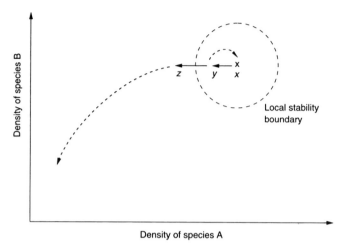

Fig. 2.3 Density of species A plotted against the density of species B and illustration of local stability boundaries by displacement from equilibrium (*x*) to position *y* or *z*.

Experimental investigation of stability over an ecologically realistic range of densities will reveal the global stability of the system.

Alternatively, one can look at the dynamics of a population or set of species over time, perturb them to different degrees at certain times and see whether they return to the same (apparent) equilibrium. For example, in Fig. 2.4, increase above or decrease below the steady state reveals the population to be locally stable between the densities *x* and *y*. These perturbations are much easier to undertake on a computer than in the field but the possibility does exist for such examinations of stability. Indeed, they may occur as a result of natural perturbations; for example, extreme weather events such as drought or hurricane.

As indicated above, ecosystem stability has been considered, like population stability, as the tendency to move to or return to a stable state. In fact, this embraces two properties of ecosystems: resistance and resilience. Resistance is a measure of the ability of an ecosystem to resist change following a disturbance such as fire or harvesting or following some change in conditions or resource supply. It is usually assessed in terms of the size of the response made to the disturbance or change. Resilience is a measure of the speed with which an ecosystem recovers after a disturbance and returns to a steady state. The effects of fire provide a good illustration of the two terms. Thus northern coniferous forest (taiga) burns easily in summer when conditions are dry, so it has a low resistance to fire. However, because some components such as black spruce (*Picea mariana*) are adapted to fire (e.g. causing the release of seeds from cones) and because fire releases nutrients from biomass and litter layers, a rapid and predictable secondary succession may

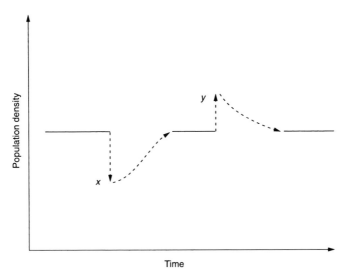

Fig. 2.4 Perturbation of a population away from equilibrium by reduction to density *x* or increase to density *y*. In both cases the population returns to the equilibrium, showing it to be locally stable.

follow fire: the system has high resilience after fire. The same arguments can be made for Californian and Mediterranean ecosystems. The idea of resilience was introduced by C.S. Holling in a seminal paper in 1973. Measures of return rates in population studies (Sibly et al. 2007) mirror the ideas of resilience in ecosystems.

2.2 Simple and complex models

Before we begin constructing our first models it is appropriate to pause and think about the rationale of model construction. The complexity of ecology and evolution provide both their fascination and frustration. We are faced with a myriad of species interacting with a variety of abiotic factors, both of which vary in time and space. How then can we begin to model these systems? There are two extremes of approach which have been described by various authors; for example, Maynard Smith's (1974) distinction between practical 'simulations' for particular cases and general 'models', May's (1973a) distinction, following Holling (1966), between detailed 'tactical' models and general strategic models and Levins' (1966) 'contradictory desiderata of generality, realism and precision'.

At the 'tactical' end of the spectrum we attempt to measure all the relevant factors and determine how they interact with the target system, such as a population. For example, in producing a model of change in plant numbers with time we might find that the plants are affected by 12 factors, such as summer rainfall, winter temperatures and levels of herbivory. This information is obtained through field observations and field or laboratory experiments. All the information is combined into a computer program, initial conditions are set (e.g. the number of plants at time 1), values for the different factors entered (e.g. the amount of summer rainfall) and the model run. The output of the model, in this case the number of plants at time t, is then revealed after different periods of time. This is a classic simulation exercise which has become feasible and easy to execute with high computer processing speeds and wide availability of appropriate software.

Now comes the tricky part. We have produced a realistic model in the sense that it mimics closely what we believe is happening in the field. However, we do not really know why it produces a certain answer. The model is intractable (and perhaps unpredictable) owing to its complexity. Tweaking a variable such as rainfall may radically change the output but we may not know why. In other words we have produced a black box which receives a set of variables and generates numbers that vary in time and space. One value of such a model is that it can speed up natural processes so that we do not have to wait 100 years to see how the plant population will (possibly) change, assuming factors remain the same or change in a predictable manner. To get closer to the mechanism(s) in these types of model we have two options. The first is to alter the variables systematically and see how the output responds. This is perhaps best undertaken after the second option, which is to strip the model down to its statistically significant components. You will recall from Chapter 1 that one feature of multiple regression is the removal of non-significant explanatory variables. This will include removing explanatory variables that are correlated. Multiple regression is one example of a set of statistical methods which allow the removal of non-significant terms, resulting in the simplest realistic model (often referred to as the minimal adequate model, especially in connection with particular statistical applications). These methods are consistent with the guiding principle of parsimony which states that the simplest explanation is the best one (Occam's razor). This principle is relevant to all branches of science. The identification of a common third and hidden variable in the power functions in Chapter 1 follows the principle of parsimony as it replaces two variables by one. In evolutionary biology parsimony has been of fundamental importance in the construction of phylogenetic trees.

From the 'strategic' end of the spectrum we can create a model which is so simple that it is known to be unrealistic. What is the value in such an approach? Here the objective is rather different to the simplest realistic model generated from the tactical end. We are using mathematical modelling as a way of formalizing generalizations about the study system. The model is not derived out of consideration of one particular example. It can be argued that strategic models are the most important types of model as they lie at the core of realistic models. If we do not understand the mode of operation of strategic models then we can never understand why the particular realistic models do what they do. From a mathematical perspective, strategic models are often designed so that their properties can be revealed through analytical solutions of the underlying equations. For example, the stability boundaries of some simple mathematical models can be determined by manipulation of the equations whereas more complex models cannot be solved in this manner. Examples of these types of solution will be given later in the book.

In the light of this discussion we will begin by exploring the properties of some simple strategic models. In fact the following could be argued as strategic models arising from realistic considerations, encompassing the best of the strategic and tactical approaches.

2.3 Density-independent population dynamics

A species with population dynamics that are relatively easy to model is one that reproduces and then dies in the same year. Certain insect species and annual plants fall into this category.

Consider populations of a hypothetical annual plant. The seed germinates in spring, the seedlings grow in the summer and reach a size for flowering and seed set in late summer. The seed are produced and over-winter in the soil. The life cycle is then repeated (Fig. 2.5). There are many variations on this theme but this is a good starting point. We will assume that the population is closed, meaning that there is no immigration or emigration.

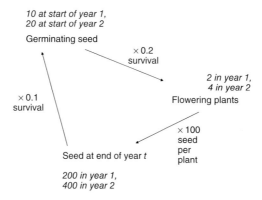

Fig. 2.5 Life cycle of an annual plant, showing change from germinating seed in spring of year *t*, through flowering and seed production in the same year. The survival and fecundity values associated with changes from one stage to the next are shown next to the arrows. The italicized values are an example of the change in numbers, starting with 10 germinating seeds.

To model the temporal dynamics of populations of this annual plant we need to estimate survival and fecundity values at different stages in the life cycle (Fig. 2.5). Assume we start with 10 seeds germinating in spring. Of these, just two survive to flower and set seed. Therefore the fraction of germinating seed surviving to flowering plant and seed set is 0.2. We will see later that these survival values can be divided further into different ages or stages of the plant (or other study organism). The fecundity of the plant is given by the average number of seeds per plant. We will take this to be 100. To take us round the life cycle to the germinating seed we now need to know the fraction of seed surviving over winter. This will be taken as 0.1. An equation can now be written for the number of germinating seed next year as a function of numbers this year:

Number of germinating seed next year = number of seed germinating this year × fraction surviving to seed set (0.2) × average number of seeds produced (100) × fraction surviving over winter (0.1)

If number of seed germinating this year is replaced by N_t (number at time t) and number of seed germinating next year is replaced by N_{t+1} (number at time $t+1$) then we can replace the above expression with a simple algebraic expression. Note that we are assuming that the survival and fecundity rates are constant between years, which allow them to be combined as one parameter:

$$N_{t+1} = N_t \times 0.2 \times 100 \times 0.1$$

$$N_{t+1} = 2N_t \tag{2.1}$$

The two survival fractions combine to give an overall value of fraction of germinating seed surviving to the equivalent point after one generation (in this case the value is 0.02). We could have done the same thing starting at a different point in the life cycle. This overall survival is then multiplied by the fecundity to give an overall measure of the change in numbers from one generation to the next. In this example the value is 2, so that the population doubles in size each year.

In mathematical models of temporal change there are two ways of representing time, which have important implications for the methods used in the modelling and the outputs of the models. In the first case, time may be considered as continuous, so that, in theory, it can be divided up into smaller and smaller units. In the second case, time is considered to be discrete in units of, for example, years. The first case is appropriate to populations of individuals with asynchronous and continuous reproduction such as human populations, whereas the second is appropriate to populations with seasonal or otherwise synchronized reproduction. The population of annual plants treated here fall into the discrete time category. The subscripts t and $t+1$ show that we are dealing with a discrete time process, with units of years, due to the fact that reproduction is annual. Such processes are modelled with difference equations (also known as recurrence equations) which relate events at one time point to those of previous time (variable) points. Equation 2.1 is an example of a difference equation. Difference equations could also be used to link different points in space. Continuous processes are modelled with differential equations. The hypothetical extremes of discrete and continuous time are not always encountered or obvious in the field. Most environments have some form of seasonality, even in tropical habitats. Within the breeding season, there may be one reproduction period or there may be several, possibly overlapping, periods of reproduction. In the latter example a combination of difference and differential equations may be required.

The model in equation 2.1 can be represented as a general algebraic equation (see equation 2.2, below), by letting λ equal the average overall survival value multiplied by the fecundity. λ is known as the finite rate of population change. This value was referred to by May (1981) as the 'multiplicative growth factor per generation'. We will see later an equivalent term representing the change in numbers of species or other taxonomic unit during evolution. In the example in equation 2.2, λ is equivalent to the number of germinating seed produced in year $t + 1$ for every germinating seed in year t (N_{t+1} divided by N_t). λ could also give a measure of the number of flowering plants in year $t + 1$ relative to the number in year t. Thus, in different versions of equation 2.2, λ can be used (and will take the same value) for any stage of the life cycle as long as it is expressed relative to the same stage in the previous cycle and the survival and fecundity values remain constant. Numbers of seeds or other stages are often represented as densities; for example, numbers per unit area.

$$N_{t+1} = \lambda N_t \tag{2.2}$$

If the survival and fecundity values remain constant the population will change by a multiple of λ every year. The population will increase if $\lambda > 1$ and decrease if $\lambda < 1$ (Fig. 2.6). At values of $\lambda > 1$ we see that increase is geometric or exponential (as expected because the values at time t are multiplied by a fixed amount). Note the similarity of the output in equation 2.2 to the exponential functions shown in Chapter 1 (Fig. 1.10). Equation 2.2 is an example of a density-independent model because the value of λ does not change with population density. The Fibonacci model discussed in Chapter 1 is also an example of a density-independent model.

The serious limitation of density-independent dynamics is that they predict an unrealistic world either eventually covered in one species (when $\lambda > 1$) or without a given species (when $\lambda < 1$). There is also the possibility of no change in population size if the death rates are exactly balanced by the birth rates: $\lambda = 1$. This is also unrealistic because birth and death rates have to be exactly matched or balanced by immigration/emigration for an indefinite period of time. This is the model of the ball balanced on the pin-head (Fig.

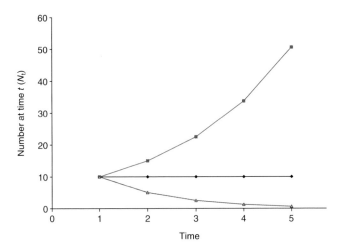

Fig. 2.6 Density-independent dynamics generated by equation 2.2 with different values of λ (top line $\lambda = 1.5$, middle line $\lambda = 1$ and bottom line $\lambda = 0.5$).

2.2). However, equation 2.2 may accurately predict dynamics over a short period of time, when the assumptions of constant rates of survival and fecundity will hold. This is likely to occur at relatively low population densities, such as when an annual plant species is colonizing a recently ploughed field. In Chapter 5 we will see how to model systems to achieve a more realistic process of stability; that is, the model of the ball in the cup (Fig. 2.2).

2.4 Density-independent growth in numbers of lineages

Just as populations increase or decrease in the numbers of individuals with time, so clades will change in the numbers of species or other taxon with time. A clade is defined as all the descendants of a common ancestor; that is, it is a monophyletic group. It is also usual to describe the number of lineages in a clade, with lineages either branching (origination events) or becoming extinct. The temporal dynamics of populations and clades have much in common in terms of modelling. Understanding the temporal dynamics of clades allows us to address some fundamental questions in evolution. For example, we may ask whether rates of evolution change significantly with events such as the end of the Cretaceous or the more recent ice ages of the Pleistocene. Formerly this type of question could be answered only with reference to the numbers of fossil types. With the development of (molecular) phylogenetic and molecular clock methods the potential for answering these questions is greatly improved.

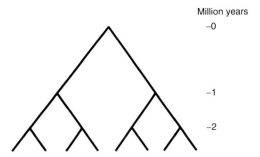

Fig. 2.7 Diversification of a clade with dichotomous branching of taxa every million years.

Imagine a clade in the early stages of diversification. For convenience we will assume time units of millions of years (Myr) and that each end point is a species. At time point 1 there are two species in existence (Fig. 2.7). During the next million years (up to 2 Myr after the clade has existed) each of the existing two species splits to form two new species; that is, four in total. If this process continues at the same rate during subsequent million-year periods the numbers will increase geometrically: 2, 4, 8, 16 and so on. This scenario does not assume any extinction events. We can model extinction in a simple manner by assuming that a certain fraction of taxa go extinct during each time period – for example 0.25 – and therefore 0.75 survive. This model is identical to the density-independent population model (equation 2.2). In this case:

$$N_{t+1} = 2 \times 0.75 \times N_t$$

$$N_{t+1} = 1.5 N_t$$

where N_t is the number of species at the end of a given time period. You will see that we also have a parameter equivalent to λ, the finite rate of population increase. This can be defined as a diversification rate, for which we will use the symbol R. In Chapter 3 we will discuss ways in which these processes can be modelled as probabilities (or as values sampled from a probability distribution) rather than fixed values.

In Chapter 1 we saw that equations such as 2.1 and 2.2, in which terms are multiplied, can be log-transformed to produce an additive model. A log-transformation of the diversification equivalent of equations 2.1 or 2.2 yields:

$$\log N_{t+1} = \log R + \log N_t \tag{2.3}$$

Note that, for a given base, $\log(ab) = \log(a) + \log(b)$. As R represents originations ('births', B) of lineages multiplied by the fraction not going extinct (S, surviving) we can write $R = BS$. Evolutionary biologists are interested in the extinction rate (E) so we can replace S with $1 - E$:

$$R = B(1 - E)$$

If R is replaced by $B(1 - E)$ in equation 2.3 we have:

$$\log N_{t+1} = \log(B(1 - E)) + \log N_t$$

which can be rewritten as:

$$\log N_{t+1} = \log(B) + \log(1 - E) + \log N_t \tag{2.4}$$

The additive components of origination and survival are clearly shown in the log-transformed equation 2.4. A second way of writing equations such as 2.1 and 2.2 is to consider the relationship between N and t. For example, after two time periods we can write an equation for N_{t+2}:

$$N_{t+2} = R^2 N_t$$

More generally, after d time periods:

$$N_{t+d} = R^d N_t$$

If N_t is the number at time 0 and therefore equal to 1 we are left with:

$$N_d = R^d \tag{2.5}$$

Using this equation we only need to know the extant number of lineages (N_d) and the total time period over which they have been in existence (d) to estimate the diversification rate. Later we will consider how to describe these processes in continuous time.

The branching patterns of real clades are far more complex than that given in Fig. 2.7. Phylogenetic trees (phylogenies) which describe the relationship between different extant or fossil groups have been determined for many clades (some examples are given in Fig. 2.8). Phylogenies are constructed by comparing the character sets of different organisms and determining their similarity. Guiding principles of phylogenetic (re)construction include parsimony, which produces a hypothesis of the simplest way of arranging the organisms by minimizing the number of transitions between character states. The statistical methods of phylogeny construction could fill a book on their own and so are not treated here.

In addition to realization of the overall structure or topology of the phylogeny, the lengths of branches can be calibrated using molecular clock techniques. Again, there is a large literature on the methods, which include the use of multiple calibration points. For our purposes, we only need to know that phylogenies can provide details of branch length and that this is proportional to time. Knowledge of the branching pattern and branch lengths means that we can plot the change in numbers through time. As an example consider the amphibian phylogeny shown in Fig. 2.8a. At time 0 (before 368.8 Myr ago) there is one branch (Table 2.1). At 368.8 Myr ago this is hypothesized to have branched into two lineages. At 357.8 Myr ago one of the branches again split to yield a total of three lineages, and so on. Sometimes the split will produce three or more branches but generally phylogenies are

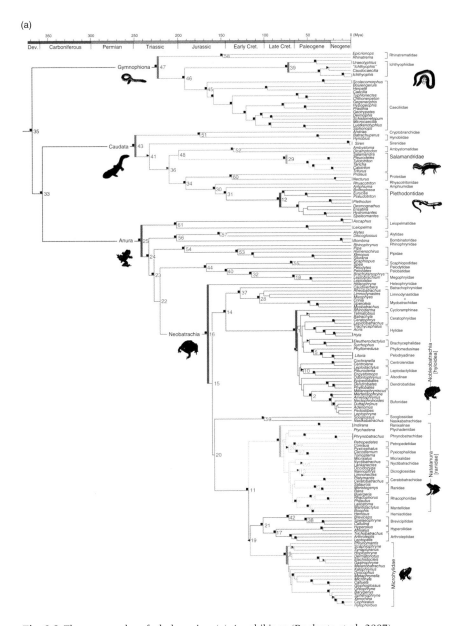

Fig. 2.8 Three examples of phylogenies. (a) Amphibians (Roelants et al. 2007).
(b) Angiosperms (see Angiosperm Phylogeny website, www.mobot.org/MOBOT/research/
APweb/). (c) Primates including fossil groups (Seiffert et al. 2005).

Continued

(b)

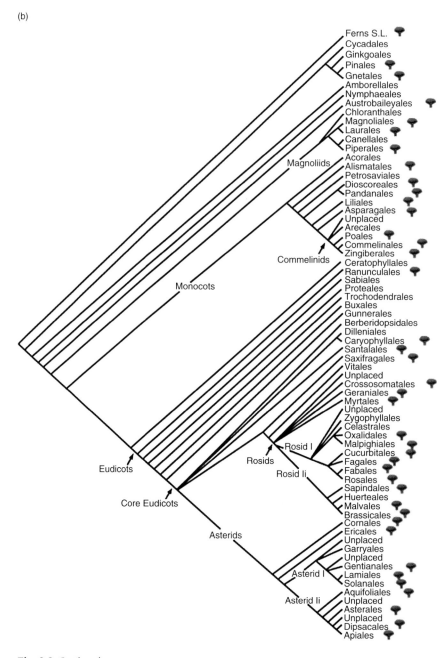

Fig. 2.8 *Continued*

(c)

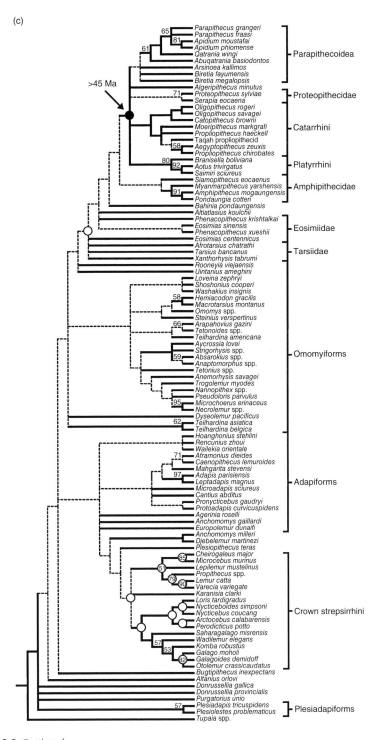

Fig. 2.8 *Continued*

Table 2.1 List of amphibian phylogeny node ages in order of decreasing age and cumulative number of lineages (Roelants et al. 2007). Only the first 12 nodes (branching points) are given. The log of the number of lineages is used in consideration of the diversification rate (Fig. 2.9).

Node ages (Myr ago)	Cumulative number of lineages	ln (number of lineages)
368.8	2	0.693147
357.8	3	1.098612
248.7	4	1.386294
242.5	5	1.609438
233.9	6	1.791759
232.2	7	1.94591
228.8	8	2.079442
226.4	9	2.197225
222	10	2.302585
209.8	11	2.397895
202.9	12	2.484907
202.5	13	2.564949

resolved to produce just dichotomous branches. A straightforward way of summarizing such data is to list the node (branching-point) ages and use this to generate a cumulative number of nodes with time (Table 2.1). As this phylogeny is based on extant lineages, it underestimates the level of extinction. However, it does represent a valuable description of temporal changes in lineages leading to extant groups. An example of a phylogeny which includes extinct groups is given in Fig. 2.8c.

Often we are interested in deviation from a null model of constant diversification. One way of exploring this is to plot the log of the number of lineages (N) against time (Fig. 2.9). Because the process is multiplicative, constant diversification should produce a linear increase in the log of number of lineages with time (see also section 1.2). Later we will see that there are good reasons for expecting systematic deviation from a linear fit.

A linear fit through all the data (Fig. 2.9) gives the following regression equation:

$$\ln N = 0.0136t + 5.2882 \tag{2.6}$$

Note that time before present is given a negative value so that the rate of increase is positive. The equivalent equation using raw numbers (without natural-log transformation) is:

$$N = e^{5.2882+0.0136t}$$
$$N = e^{5.2882}e^{0.0136t}$$
$$N = 197.99e^{0.0136t} \tag{2.7}$$

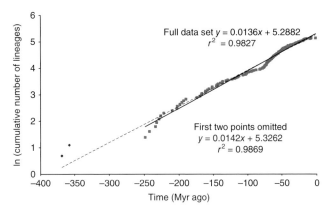

Fig. 2.9 Natural log of the number of amphibian lineages against time, based on the phylogeny in Fig. 2.8a.

The term $e^{0.0136}$ gives the average rate of increase of lineages per unit time. This is equivalent to 1.0137, which is the average rate of multiplication of lineages every million years. Another approach to estimating the average rate of diversification is to take the known number of extant species and use equation 2.5. With 6009 extant species and a start point of 368.8 Myr ago we have:

$$6009 = R^{368.8}$$

In other words R is the 368.8th root of 6009! This can be written as $6009^{1/368.8}$ or $6009^{0.00271}$ and has a value of 1.023873. So, starting with one lineage 368.8 Myr ago, multiplication every million years by 1.023873 gives approximately 6009 at the present day. Note that with so many multiplications rounding error is important here. More sophisticated measures of diversification rate assume certain levels of extinction. Using these amphibian data Roelants et al. (2007) estimated the diversification rate with no extinction as 1.0217 and with a very high extinction rate of 0.95 times that of birth rate they estimated the diversification rate as 1.0154.

It is clear from Fig. 2.9 that there are some major deviations in the values around the regression (which we could explore by examining the residuals of the regression). For example, the numbers build from a low value at approximately 250 Myr ago to a maximum at about 190 Myr ago. There is also a dip at about 75 Myr ago. Analyses of deviations from constant diversification are of considerable interest as they may indicate periods of high origination or high extinction. If such patterns are found across different clades then this may constitute evidence for global patterns of high extinction; for example, at the Cretaceous/Palaeogene boundary 65.5 Myr ago.

The point at which the line cuts the *x* axis will be the point at which ln (number of lineages) = 0 and so the number of lineages = 1. In other words this would be an estimate of the time of the ancestral lineage. Of course we already have this value from the molecular phylogeny, but it is interesting to see if the regression agrees with this by removing the early values. In Fig. 2.9 there may be an effect of the two very early branches on this estimate (recall the effect of points a long distance from the midpoint of the regression). This can be explored by undertaking the regression with these two points omitted. The resulting regression equation is:

$$\ln(\text{numbers of lineages}) = 0.0142t + 5.3262$$

This can be rearranged with ln (number of lineages) = 0 to give *t*:

$$t = -5.3262/0.0142 = -375.1$$

Therefore this predicts a start point of 375.1 Myr ago.

An important systematic deviation in plots of ln (lineages) against time for molecular phylogenies is due to changes in the extinction rate (Nee 2006). As molecular phylogenies are based on extant species they tend to underestimate extinction, especially as graphs such as Fig. 2.9 approach the present, where the slope is more likely to approach the origination value (*b*; Fig. 2.10). The birth/death process described in the legend refers to a different model from that used above and one which we will consider later, although the principle of change in slope is still relevant. In Chapter 5 we will discover other sources of deviation from the linear plot in Fig. 2.10.

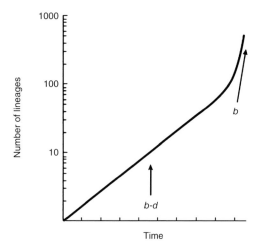

Fig. 2.10 The expected cumulative increase in the logarithm red correction of number of lineages in a molecular phylogeny growing according to a birth (*b*)/death (*d*) process (Nee 2006). The birth/death process is described in Chapter 3.

2.5 Population dynamics and diversification in continuous time

What of organisms whose life cycles do not fit the simple assumptions of annual plants or insects? From the point of view of population dynamics there are three important differences between long-lived organisms and annual organisms. First, the former may begin reproducing after more than 1 year. Second, they may survive after reproduction and possibly reproduce again. The dynamics of such populations needs to be described with respect to particular ages or stages of the population and will be the subject of Chapter 4. Finally, in populations with overlapping generations, reproduction may not be discrete or synchronous. Individuals in human populations, for example, do not synchronize their reproduction! In this case we need to think of reproduction as a continuous rather than a discrete process. Similarly, in describing the diversification of clades it may be more appropriate to describe change as continuous.

In section 2.3 we considered an equation in discrete time ($N_{t+1} = \lambda N_t$) to represent density-independent population change. In section 2.4, the use of an analogous discrete time model was applied to diversification rates. If reproduction or diversification in a large clade is continuous then the difference between t and $t + 1$ is vanishingly small and therefore change is continuous and described by differential equations. In the following example we will discuss population change but the same ideas apply to diversification rates.

In the discrete-time model it was found that population change was geometric in form (Fig. 2.11a). Now consider a continuously reproducing population. For a description of continuous geometric population change the separate points in Fig. 2.11a need to be replaced by a smooth curve (Figs 2.11b–2.11d).

If a population is changing geometrically (exponentially) then the curve (Fig. 2.11c or 2.11d) can be described by an equation which we have met for the discrete-time process (equation 2.7):

$$N_t = N_0 e^{rt} \tag{2.8}$$

where N_t is the population size at time t and N_0 is the initial population size. Because equation 2.8 is describing continuous change, t can take any value. At each point on the smooth exponential curve it is possible to determine the rate of population change by differentiation. This is equivalent to the gradient of the tangent at that point. Tangents represent linear rates of change at one point on a nonlinear curve. Differentiation, which is one branch of calculus, provides a way of finding the gradient at a given point on a curve produced by a known function. Differentiation therefore provides a means of determining the rate of change of one variable in response to another. Moreover, whereas drawing a tangent is only an approximate way of finding a rate of change at a particular point, differentiation provides a precise value. Differentiation essentially provides a method of quantifying

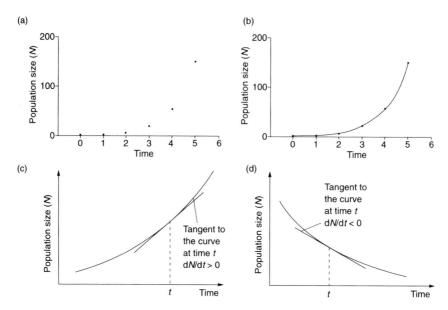

Fig. 2.11 (a) Geometric or exponential growth in discrete time. (b) Geometric growth in continuous time. (c, d) Geometric growth in continuous time showing tangent to curve for (c) positive and (d) negative rates of change.

very small changes in the dependent variable with very small changes in the independent variable at a given value of the independent variable.

In the terminology of differentiation, the rate of population change at a given time t (Fig. 2.11c or 2.11d) is referred to as the derivative of N_t with respect to t and is written as dN/dt (this is described in speech as 'dN by dt'). When dN/dt is positive the population is increasing with increasing time (Fig. 2.11c); conversely when dN/dt is negative the population is decreasing with time (Fig. 2.11d) and when $dN/dt = 0$ there is no change in population size. There are various rules for differentiating different functions. We will not go through all these here but it is helpful to know a couple of them. One you have already met in Chapter 1 is if $y = e^x$ then $dy/dx = e^x$. This is a special property of the exponential curve in which the gradient of the curve at point x is equal to the value of y at that point x. The derivative of the function $y = ax^n$ is given by:

$$dy/dx = anx^{n-1}$$

For example, for the quadratic equation $y = 3x^2 + 2x + 4$ we can treat the three added components separately (note that 4 is lost because we are multiplying by a power of 0):

$$dy/dx = 3.2x^1 + 2.1x^0 + 4.0$$

$$dy/dx = 6x + 2$$

The value of the gradient at, for example, $x = 2$ could then be determined ($dy/dx = 14$). A useful application of differentiation is when we wish to examine maxima and minima in curves of population change. Maxima and minima are defined as points where no change occurs (therefore $dy/dx = 0$). However, they differ in the way that dy/dx changes with x (Fig. 2.12). For example, consider increasing values of x approaching and passing a maximum value. Before the maximum, the value of dy/dx is positive. These positive values reduce towards zero at the maximum value and thereafter become increasingly negative. Therefore dy/dx declines from high positive to zero to negative values with increasing values of x across a maximum. The rate of change of dy/dx with x is known as the second derivative (the derivative of the derivative, written as d^2y/dx^2) and can be used to distinguish between maxima, minima and points of inflexion (Fig. 2.12).

Returning to equation 2.8 we can now see that the population increase at a particular time is found by differentiating N_t with respect to t:

$$dN/dt = rN_0e^{rt} \tag{2.9}$$

By substituting $N_t = N_0e^{rt}$ into equation 2.9 we obtain:

$$dN/dt = rN_t \tag{2.10}$$

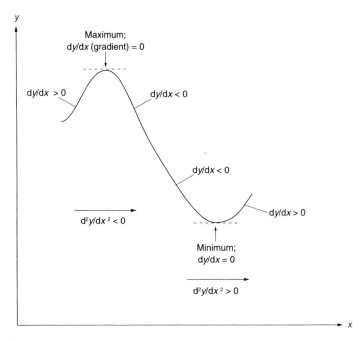

Fig. 2.12 Illustration of maximum and minimum values of a function $y = f(x)$.

Equation 2.10 shows that the rate of population change at time t (dN/dt) is equal to the population size at that time (N_t) multiplied by r. The parameter r is referred to as the intrinsic rate of population change. Despite its name, there is nothing intrinsic about r. Its value will change as rates of birth, death and dispersal change. So it is better to reserve the term intrinsic for the maximum value of r (r_m) which occurs under optimum conditions of temperature, light, food supply and so on. The actual instantaneous rate of increase, r, will always be lower than r_m and will vary with time for a particular species. r or r_m is also called the Malthusian parameter.

In a closed population r represents the difference between the birth and death rates per individual. Thus r can be replaced by $b - d$, where b is the instantaneous birth rate and d is the instantaneous death rate. The same notation can be applied to origination and extinction of lineages in clades (see Fig. 2.10). If the birth rate exceeds the death rate then $b > d$ and so $r > 0$, meaning that dN/dt is positive and the population will increase in size. Conversely, if $b < d$ or $r < 0$ then the population will decrease in size ($dN/dt < 0$). There is also an unstable steady state given by $b = d$ (analogous to $\lambda = 1$), so that $b - d = 0$ and there is no change in population size; that is, $dN/dt = 0$.

To illustrate the estimation of r we will use the example of population change in the USA from 1790 to 1910. Although these data were presented by Pearl and Reed (1920) to illustrate a different point, it is interesting to use them here to contrast with their analysis, which we will discuss in Chapter 5. To estimate the parameter r we linearize equation 2.8 by taking the natural log of both sides:

$$\ln(N_t) = \ln(N_0) + rt$$

r can be estimated by linear regression of $\ln(N_t)$ against t, giving a value of 0.027 (Fig. 2.13a). The linear fit is apparently very good, explaining 99.5% of the variance. However, the pattern of residuals around the regression suggests that extrapolation of the linear model beyond 1910 may not be appropriate (Fig. 2.13b). If the linear model was appropriate we would expect an even scatter of points around the line. In this example the value of r is estimated over a period of time when high levels of immigration were occurring in the USA and therefore is likely to be higher than r estimated for a closed population.

Finally, let us consider the relationship between the differential equation $dN/dt = rN_t$ (equation 2.10) and the difference equation $N_{t+1} = \lambda N_t$ (equation 2.2). Both of these equations describe geometric or exponential population change; the first in continuous time and the second in discrete time. The rate of population change is given by r and λ respectively. But what is the relationship between these two parameters? Consider values of population density at two consecutive points in time. The differential equation is derived from $N_t = N_0 e^{rt}$ (equation 2.8). With $N_0 = 1$, at $t = 1$ $N_1 = e^r$ and at $t = 2$ $N_2 = e^{2r}$. (Remember that this continuous model can have values between $t = 1$ and

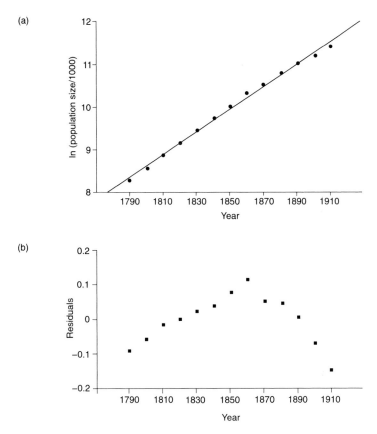

Fig. 2.13 (a) Growth of the human population in the USA from 1790 to 1910 (data in Pearl and Reed 1920). Data plotted as ln (population size/1000) against year (t). For example, in 1910 the population was estimated as 19 970 000. (b) Residuals of the linear regression in (a).

2, unlike the discrete-time model.) Dividing N at time 2 by N at time 1 gives $e^{2r}/e^r = e^r$ (note that $e^{2r} = e^r \times e^r$ and that any value of N_0 could have been used as it would cancel out here). So, the continuously growing population increases by e^r between all consecutive integer time values. By comparison, the difference equation describes a change between two consecutive integer time values as $Nt + 1/Nt = \lambda$. Thus λ is seen to be equal to e^r, or $\ln \lambda = r$.

CHAPTER 3

Stochastic models

3.1 Probability distributions

In a deterministic world everything would be predictable. If speciation rates were deterministic we would be able to predict exactly the number of species at time $t+1$ given the numbers of species at time t and a knowledge of the underlying processes governing speciation. This notion of a deterministic and therefore predictable world is upset by two important phenomena. First, and most obviously, many environmental phenomena are not deterministic! Randomly occurring, or generally unpredictable, events make an important contribution to ecological and evolutionary processes. In these cases we use the term *stochastic*. The second issue is that, even when processes are deterministic, the results may appear to be random. Thus chaotic phenomena, generated by strictly deterministic processes, produce apparently random output (Chapter 5).

Many unpredictable phenomena have a set of possible outcomes. In some cases there may be only two possibilities, such as whether or not it rains on a given day. Similarly, we may consider whether or not a species will go extinct in a given time period. Other phenomena will have more than two outcomes. The probability of a particular outcome can be determined based on considerations of different temporal or spatial scales. The probability that it rains tomorrow could be judged on how many days it has rained in the last month; for example, 28 out of 30 days. We might wish to contrast this probability (28/30) with that of equal probability (1/2) that it rains or does not. The much higher probability of rain during that month may indicate that we are in a wet season or simply an area of high rainfall.

Let us assume that the probability that it rains on a given day is 0.75 based on past events over several years. This might suggest that we can predict the weather (!) but we cannot be certain whether it will rain on a given day. In fact, we have a probability of 0.75 that it rains on a given day and 0.25 that it does not rain. Assuming that rain today is not affected by rain yesterday – that is, that rain on a given day is independent of rain on another day (this is an important assumption and one we will modify later) – we can generate a binomial distribution of events as follows.

44

Over 2 days, the probability that it rains on both days is 0.75×0.75; the probability that it rains on one day is 0.75×0.25 ($\times 2$ as it can happen in either order) and the probability that it does not rain on both days is 0.25×0.25. The distribution of probabilities that it will rain or not over different numbers of days builds in the following manner:

1 day: 0.25 or 0.75;

2 days: 0.25×0.25 (no rain), $2 \times 0.75 \times 0.25$ (rain on 1 day), 0.75×0.75 (rain on both days);

3 days: $0.25 \times 0.25 \times 0.25$ (no rain), $3 \times 0.75 \times 0.25 \times 0.25$ (rain on 1 day), $3 \times 0.75 \times 0.75 \times 0.25$ (rain on 2 days), $0.75 \times 0.75 \times 0.75$ (rain on 3 days).

In each case the probabilities sum to 1 (day 1, $0.25 + 0.75 = 1$; day 2, $0.0625 + 0.375 + 0.5625 = 1$ and so on). The distribution of probabilities rapidly becomes complex as the number of days increases, even though we are only dealing with two events (rain or not). For this reason statisticians have devised shorthand algebra to summarize the probability distributions. In the case of the binomial distribution, let p equal the probability of one event and q equal the probability of the other ($p + q = 1$). If n is the number of days we can determine the probabilities of rain on zero up to n days by expansion of $(p + q)^n$; the following are expansions for $n = 1$–4:

$$p + q$$
$$p^2 + 2pq + q^2$$
$$p^3 + 3p^2q + 3q^2p + q^3$$
$$p^4 + 4p^3q + 6p^2q^2 + 4pq^3 + q^4$$

Notice that the coefficients (the number of p and/or q combinations) increases in a predictable manner, this is known as Pascal's triangle:

$$1\ 1$$
$$1\ 2\ 1$$
$$1\ 3\ 3\ 1$$
$$1\ 4\ 6\ 4\ 1$$

Here each coefficient is the sum of the above two in the previous row. The coefficient can be generalized by a formula, $n!/(s!(n-s)!)$ where s is the number of events with probability p and $n!$ is n factorial. n factorial means that the integers from n to 1 are multiplied; for example, 3! is $3 \times 2 \times 1 = 6$. So, the coefficient for three rainy days out of four is:

$$4!/(3!(4-3)!) = 4 \times 3 \times 2 \times 1/(3 \times 2 \times 1(4-3)!) = 4$$

$p = q = 0.5$ is an example of a uniform distribution, which also occurs for more than two outcomes; for example, in the roll of a die where the values 1–6 have an equal probability of occurring (1/6).

A probability density function (pdf) is a set of mathematical statements that tell us the probability that a variable will take a given value. The sum of probabilities in a pdf is 1. Pdfs can be discrete, such as the binomial example

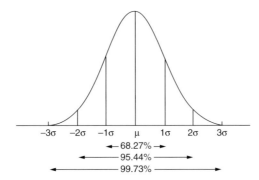

Fig. 3.1 Areas under the normal probability density function, showing the percentage of events occurring between one, two or three standard deviations (σ) either side of the mean (μ).

or the uniform distribution of rolls of a die, or continuous, such as the normal distribution (Fig. 3.1). For a continuous distribution we cannot say that a variable will have a certain value but instead we say that it can lie between different values with a certain probability. For a normal distribution, the probability that a variable will lie between one standard deviation either side of the mean is 0.6827 whereas the probability of it lying within two standard deviations either side of the mean is 0.9544 (Fig. 3.1). So, if a variable is normally distributed we expect 68.27% of the values to lie within one standard deviation of the mean.

A process by which events occur at random in space or time is known as a Poisson process. The distribution of those events – the number of events occurring per unit of time or space – is described by the Poisson distribution. The Poisson distribution is an example of a discrete pdf as it is concerned with counts of events. A Poisson process is recognized by its properties of homogeneity and independence. By homogeneity, we mean that the probability of an event occurring per unit time or space remains constant. The assumptions of independence and homogeneity mean that the Poisson distribution is a useful null model in ecology and evolution. For example, we might hypothesize that the distribution of plants in a field are clumped or aggregated because the plant reproduces asexually from its roots. This hypothesis can be tested against the null model of random distribution in space which can be modelled with the Poisson distribution. If the mean number of plants per square metre is given as x, then the terms of the Poisson distribution are:

$$e^{-x}, xe^{-x}, (x^2/2!)e^{-x}, (x^3/3!)e^{-x} \ldots (x^n/n!)e^{-x}$$

The first term gives the probability of 1 m^2 of ground containing zero plants, the second term gives the probability of 1 m^2 containing one plant and so on.

The fact that the terms can be summed to 1 means that we can determine the probability that a square metre contains at least one plant by calculating $1 - e^{-x}$. Note that the Poisson distribution is concerned with relatively rare events. In this case, it requires that the mean number of plants (x) per square metre is small compared with the maximum possible number of plants in that area. The number of samples predicted to contain 0, 1, 2, 3, etc. plants can be found by multiplying the probabilities in the Poisson terms by the total number of samples. The observed distribution can then be tested against this predicted Poisson distribution. A suitable significance test can determine whether this is just chance or a significant departure from random. Note that inspection of the data is important here as the distribution could depart from random but be regular rather than clumped. The same principles of null hypothesis testing apply to clumping in time.

An alternative to testing for clumping against a Poisson process is to find a distribution that assumes a clumped distribution. The negative binomial is an example of such a distribution, with an extra parameter k which reflects the degree of clumping. As k increases, the negative binomial approaches the Poisson distribution.

Any set of environmental dynamics is likely to be composed of deterministic and stochastic elements. A major issue in modelling is to tease apart these two elements and determine their relative importance. The regression analogy is helpful here in that we seek to quantify the relative amount of explained (deterministic) and unexplained (stochastic) variation: these two components are sometimes referred to as the signal and the noise. Just as there may be several components of the deterministic variation (as revealed by multiple regression) the unexplained variation may have several sources. In population dynamics the unexplained variance is composed of extrinsic random events (environmental stochasticity), variation between individuals in survival and fecundity, sampling error and non-significant deterministic factors. Although the overall levels of variation in survival and fecundity may be predictable, in smaller populations they combine with sampling error to generate essentially random mixes of individuals, a phenomenon termed demographic stochasticity. In this chapter we will consider how to begin modelling stochastic events.

3.2 Random walks and evolution

3.2.1 Generating a random walk

A random walk proceeds from a start point and travels a fixed distance from one point to the next in a random direction (no one direction is more likely than another; Fig. 3.2). Random walks can be in space or time. Both of these are useful in developing null models in ecology and evolution. For example,

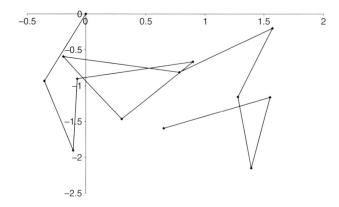

Fig. 3.2 The result of a random-walk model in two dimensions. The walk starts at (0,0) and proceeds a fixed distance of 1 unit in a random direction. Note the difference in scale of the x and y directions.

a study of foraging in insects may wish to compare observed foraging distances with those generated from a random walk. Similarly, we may contrast the evolutionary change through time in a particular character or the number of species against that expected from random.

Let us imagine a random walk as a model of change in number of species over time. We start at time 0 with 100 species. At each time step the number of species can increase or decrease. For a purely random model we would assume that the mean increase is 0 (there will be situations in which we will want to change this assumption, giving clades or populations net rates of increase or decrease). Therefore, at each step in the time series we choose a value at random from a probability distribution. An important consideration is whether the model is additive or multiplicative. This also applies to similar models of change in population size over time (Lewontin & Cohen 1969). In a simple additive model the number of species (or individuals in the case of a population model) to be added is independent of the number of species or individuals at that time. This is likely to be an oversimplification. A better model is that used in equation 2.2; that is, that the number of species or individuals in the next time period is a multiple of the number now:

$$N_{t+1} = \lambda N_t$$

If we want no mean change in size with time λ is set to 1. In the new model we include a variable ε that can take values at random:

$$N_{t+1} = \lambda N_t \varepsilon \tag{3.1}$$

As the model is multiplicative it is best to transform to log values so that increases and decreases are displayed as equal values. For example, in a multiplicative model multiplying by 10 and multiplying by 1/10 are equal but opposite. To express these multiples as the same size (magnitude) they are transformed to log values; for example, $\log_{10} 10 = 1$ and $\log_{10} 1/10 = -1$. The new model (equation 3.2) can be expressed with log values:

$$\log N_{t+1} = \log \lambda + \log N_t + \log \varepsilon \qquad (3.2)$$

Using this transformation will also help when we come to estimate parameter values with linear regression techniques. Mean increase in the multiplicative model over time is seen to be 0 when using log values.

Consider an example in which ε is a value taken at random from a normal distribution with a mean of 0 and standard deviation of 1 (Fig. 3.3a). We can use a series of these random numbers to generate a time series with different values of λ. In this case we illustrate $\lambda = 1$ (Fig. 3.3b). Notice the relatively smooth shape of this walk compared with the spiky nature of the random values (Fig. 3.3c).

The contrast between the two time series in Figs 3.3b and 3.3c can be revealed by correlation of the numbers at time $t + 1$ with those at time t (Fig. 3.4). This shows that there is a strong positive correlation for the random-walk model (Fig. 3.4a) compared with no correlation between the random time series (Fig. 3.4b). This is correct because the random numbers are drawn independently from a normal distribution whereas the speciation model (equation 3.1 or 3.2) makes N_{t+1} a function of N_t. Correlations between sets of time series data are referred to as autocorrelations. In this case we have considered an autocorrelation of lag 1; that is, a difference of one time step. Autocorrelations with lags of more than one can also be studied and may be expected to occur when different species interact (Chapter 6). In general, autocorrelation is a useful technique for starting to explore signals in a time series with some level of stochasticity.

Random walks have been used as null models in studies of change in marine fossil diversity with time (Cornette & Lieberman 2004). This study made use of Sepkoski's compendium of fossil marine genera and showed that changes in diversity over the last 540 Myr are consistent with a random walk. This does not necessarily mean that the underlying processes are stochastic, but that the net result of the processes causing change appears to be random. Indeed, the same data set has also been used to address periodicity in the fossil record. Most famously, it was the basis of Raup and Sepkoski's analysis that led to the idea of mass extinction events, such as the end-Permian and end-Cretaceous (Raup & Sepkoski 1982), which, in turn, were linked with periodicity of approximately 26 Myr and a galactic cause of extinction (Raup & Sepkoski 1984). More recently, Sepkoski's fossil data set has been reanalysed to reveal 62 and 140 Myr periodicity (Rohde & Muller 2005). Therefore, this fossil data set, spanning the entire duration of the Phanerozoic, illustrates both deterministic (periodic) and stochastic (random-walk) processes.

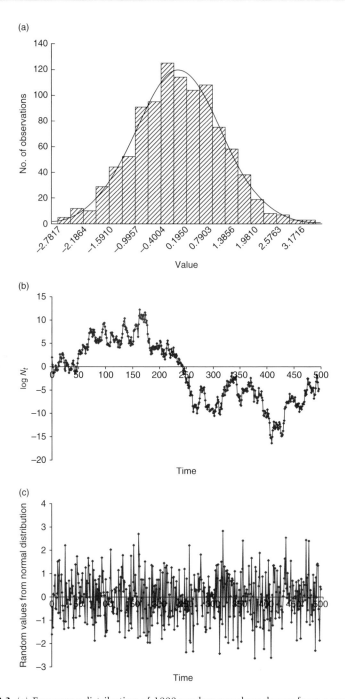

Fig. 3.3 (a) Frequency distribution of 1000 random numbers drawn from a normal distribution with mean of 0 and standard deviation of 1. The curve shows the normal density function drawn from the mean and standard deviation of the observed data (mean of −0.0002 and standard deviation of 0.991). (b) Random-walk time series with $\lambda = 1$ and equation 3.2. (c) Time series of random numbers used in (b).

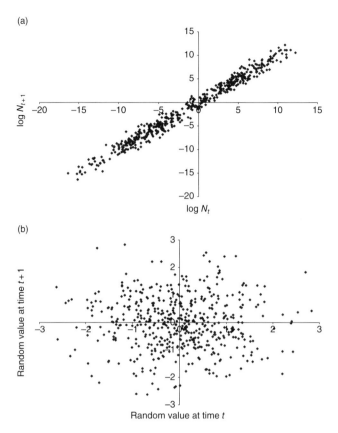

Fig. 3.4 (a) Autocorrelation between numbers at time $t + 1$ and t in the time series of Fig. 3.3b (equation 3.1 model) and (b) the random numbers used in that model (Fig. 3.3c).

3.2.2 Birth/death processes in evolution

We have a number of options in constructing models of diversification. We can use discrete or continuous time models with deterministic processes as in Chapter 2. Alternatively we can employ stochastic models in discrete or continuous time (Nee 2006). The random-walk model above is an example of a discrete time process. Running this simulation many times would produce a set of possible values of clade richness which could serve as a null model for diversification. A key difference to the deterministic model is that there is no single outcome; instead the set of possible outcomes is defined by a particular distribution. These outcomes include complete extinction of the clade. Understanding and quantifying that distribution of possibilities is an important goal in evolutionary study.

The simplest and earliest model of diversification was the pure birth model in which each species has a constant probability of producing a new species and there is no extinction. This leads to the prediction of exponential growth, as with the deterministic analogue. The pure birth process is also known as the Yule process (Yule 1924). Yule's paper is a rich source of information on the ways in which evolution can be modelled. He shows how a mathematical model of evolution can be constructed from first principles and the various predictions that can be made. This includes examples of the probability distributions that arise from iterations of all possible outcomes of the model. In Fig. 3.5 the first two steps of the simplest model are shown. p is the probability of a species producing a new species (Yule referred to these as mutations) and q is the probability of that event not happening ($p + q = 1$). Yule was interested in the distribution of species within genera. Row 1 in Fig. 3.5 shows the probability of a genus containing one species after two time intervals. The time intervals were considered to be sufficiently small that two p events were highly unlikely. Rows 2 and 3 are then two outcomes that result in two species per genus. Notice that in all cases the values of p and q are multiplied together as they are independent events. After two time steps the probabilities associated with one, two, three or four species per genus are as follows: q^2, $pq + pq^2$, $2p^2q$ and p^3 respectively.

These terms sum to 1 as required (you can check this by substituting $1 - p$ for q). Yule extended the process to a large number of time steps, demonstrating that the terms formed a geometric series. The process was also developed for different-aged clades and tested against different data sets.

A natural development of the model is to include an extinction term. The characteristics of the stochastic birth/death process are as follows (Magallon & Sanderson 2001). Speciation and extinction are assumed to occur at constant rates, b and d respectively, which produces exponentially declining or increasing diversity. The diversification rate is defined as $b - d$ whereas the relative extinction rate is d/b. The probability distribution for the number of lineages at a given time, t, is also known, as are the confidence limits for diversification rates. The fact that clades may go extinct before they are sampled is just one of several problems that face the interpretation of birth and birth/death models, especially when using phylogenies based on extant data. The example from Nee (2006) in Chapter 2 (Fig. 2.10) illustrates how, under a stochastic birth/death model, the cumulative increase in the logarithm of lineage numbers is expected to approach b with increasing time (towards the present). Rates of diversification are generally presented under different extinction scenarios, for example 0 and 0.9 (Magallon & Sanderson 2001).

3.3 Probability of population extinction

In the last section we considered extinction events with an evolutionary perspective, which generally includes long timescales and large-scale extinc-

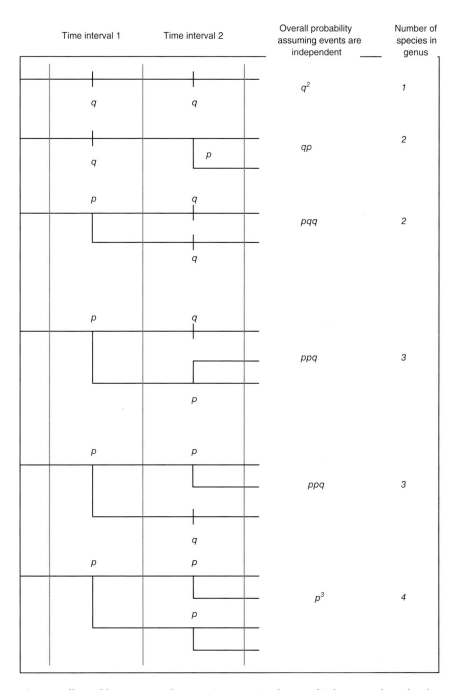

Fig. 3.5 All possible outcomes after two time steps in the pure birth process formulated by Yule (1924). Overleaf the outcomes are expressed as a tree diagram.

Continued

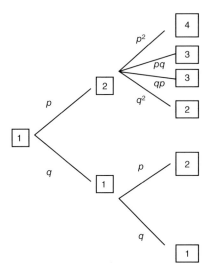

Fig. 3.5 *Continued*

tion events (not necessarily 'mass' extinctions – see the review in Bambach 2006 – but perhaps of many species). The corollary of these activities is a focus on extinction today, where activity may be aimed at smaller taxonomic units. Conservation biologists have addressed both the mechanisms underlying population extinctions and ways of assessing the likelihood of extinction (e.g. Pimm et al. 1988, Foley 1994, Holmes et al. 2007). A recurring problem in conservation biology is how to use population abundance data and population dynamics theory to evaluate which populations of which species should be protected or managed. International conservation organizations, such as the International Union for Conservation of Nature and Natural Resources (IUCN), have become increasingly interested in the application of population biology theory to conservation problems, beginning with Mace and Lande (1991). This process and its outcomes are referred to as population viability analysis (PVA). Here we will examine how modelling techniques can be used to assess the likelihood that a population will become extinct over a given time period.

Extinction may be caused by both stochastic and deterministic processes. Deterministic processes might include habitat loss or hunting (although these will be stochastic in the short term), whereas stochastic processes may include extreme weather events. To model extinction we may think of stochastic processes as reducing population size in the short term but not affecting mean population size, whereas deterministic events reduce the mean population size in a predictable manner. A combination of these two types of process can also be envisaged; that is, a population with a declining mean size that also fluctuates about the mean (Fig. 3.6).

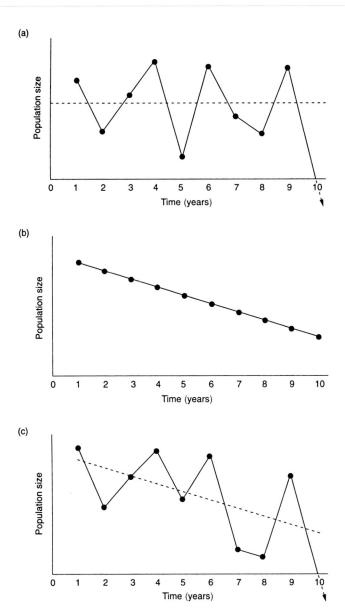

Fig. 3.6 Types of population extinction. (a) Extinction due to fluctuations around a constant mean population size (a stochastic process). (b) Extinction due to a decreasing mean population size (a deterministic process). (c) A combination of (a) and (b).

Fluctuations in population size over time may be large and unpredictable. Populations can be modelled by equation 3.1 in a similar manner to species or other taxonomic diversity through time. In a given period of time there is the possibility that a sufficiently large fluctuation will cause the population to become extinct. An estimate of the probability of extinction allows us to quantify that possibility. The random-walk model described above gives one method of modelling these processes. We need to define probabilities of extinction for populations over a given time period; for example, a population may have a 1 in 10 chance of becoming extinct in any one year.

We will apply these ideas to the density-independent model $N_{t+1} = \lambda N_t$. In Chapter 2 λ was assumed to be held constant and in equation 3.1 it was multiplied by a value drawn at random from a probability distribution, ε. In this example we will retain the stochastic element but simplify the probability distribution. As before, we are going to assume that the changes in λ are random in their operation so there are good and bad years for populations and these are entirely unpredictable in their occurrence over time. These unpredictable factors could affect λ through either the survival of different stages and/or the fecundity of individuals. If λ can take a range of values, each with a certain probability, it is no longer described by its mean, but by a pdf. In this case we will assume a discrete pdf of six values, each of which has the same probability (an example of a uniform distribution; Fig. 3.7). If we take a value of λ at random from this distribution and multiply it by an initial size at time 1 (N_1), this will give the value of N_2. N_2 is then multiplied by a new value of λ, plucked again at random from the pdf, to give N_3 and so on. The six possible values of λ can be matched into three pairs: 1/10, 10; 1/2, 2 and 3/4, 4/3. So, for example, the population has the same chance of being halved as being doubled. We might therefore expect the net change in population size to be zero; that is, that the population will fluctuate around its original level. However, there is a chance of a number of bad years in a row, which might lead to extinction. For example, five very bad years in a row, starting from an initial population size of 1, would give a population size of $1 \times (1/10)^5$.

It is necessary to set an arbitrary extinction density greater than zero because the population will not reach zero given the assumptions of the model. In fact, the population moves asymptotically towards zero. Once the extinction size (or density) is set a proportion of the population following the dynamics described by $N_{t+1} = \lambda N_t$ and the pdf in Fig. 3.7 may become extinct over a given time period. A simulation using initial population sizes of 1, 5 and 10 is shown in Table 3.1. The simulation was repeated 10 times for each initial population size and stopped after 20 time periods if the population had not become extinct. An arbitrary extinction density of 0.5 was assumed. It can be seen that increasing the initial population size from 1 through to 10 decreased the probability of population extinction and increased the mean persistence time, which is what we would expect. It may seem surprising that, with an initial population size of 1, any of these populations persisted at all. The fact that three out of 10 did persist is a consequence of the high variance in the distribution of λ.

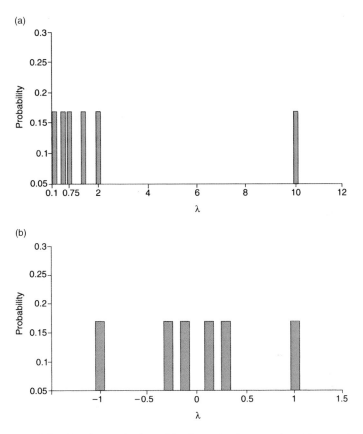

Fig. 3.7 Probability distribution for λ used in the extinction exercise: (a) raw values and (b) log-transformed values.

Table 3.1 Simulation results for a density-independent model $N_{t+1} = \lambda N_t$. Each simulation was for a maximum of 20 time periods and was replicated 10 times with the pdf illustrated in Fig. 3.7.

Initial population size	Probability of extinction	Mean persistence time (generations)
1	0.7	8.8
5	0.6	12.4
10	0.3	16.7

λ was made to follow a uniform distribution, the arithmetic mean of which is 2.45. Despite this it is not predicted that the population would increase in size if it persists. This is because we need to consider the mean effect of multiplying N_t by λ, not the mean effect of adding λ. In this example, the pdf was so simple that we could see that, on average, the mean effect would be as if λ was 1 (indeed, the values were chosen for that reason!). So, the net effect of multiplying N_t by the six λ values was to multiply by 1 (as with equation 3.1); that is, not to change the original value of N when averaged over many years. Rather than the arithmetic mean of N it is the geometric mean that is relevant here. This applies to all models of the type $N_{t+1} = \lambda N_t$. The geometric mean of a set of r numbers is found by multiplying them together and taking the rth root (recall the method of determining the diversification rate in Chapter 2). This is equivalent to taking the logarithm (to a given base) of the raw values, taking the arithmetic mean of those values and then back-transforming to get the geometric mean. As we saw with equations 3.1 and 3.2, using the logarithms of the numbers gives us an additive rather than a multiplicative model. Thus, $N_{t+1} = \lambda N_t$ is transformed to log $N_{t+1} = \log \lambda + \log N_t$. If we look at the pdf of $\log_{10} \lambda$ we see that the arithmetic mean is zero (see Fig. 3.7, where the bars are symmetrical about zero) and therefore, on average, $\log N_{t+1} = \log N_t$. Back-transforming (taking the antilog) gives the geometric mean of 1.

Models of population extinction have used these ideas to estimate probabilities of extinction and mean times to extinction (or, conversely, persistence time) for a wide range of species. Foley (1994) used the density-independent model $\ln N_{t+1} = \ln \lambda + \ln N_t$, assuming that $\ln \lambda$ was normally distributed with a variance v_r and mean 0, and that the population started at size N_0 and proceeded on a random walk with a maximum value of k and a minimum value of 0: extinction. The mean time to extinction, T_e, is then given as:

$$T_e = (2\ln(N_0)/v_r)(\ln(k) - \ln(N_0)/2)$$

This model was applied to different species, showing times to extinction of 19–237 years for five populations of the wolf (*Canis lupus*) and 1378–6107 years for six populations of the mountain lion (*Felis concolor*). It would be unwise to accept these and other estimates of T_e or probability of extinction as absolute estimates. They are based on simple models with a series of assumptions. Perhaps their most useful function is to provide an estimate of the relative likelihood of extinction, as a contribution to a population or species viability assessment, such as that of the IUCN Red List.

3.4 Extension material

It is worth commenting that this chapter on stochastic processes is the tip of a statistical iceberg of methods and results. Some key concepts and techniques that accompany these ideas include the types (colours) of noise and their relationship to autocorrelation, Markovian processes (Chapter 4), calculation of confidence intervals and maximum likelihood techniques. All of these methods require an appreciation of statistics which is generally far beyond that in a standard undergraduate course in ecology and evolution. However, there is no doubting the importance of these subjects for research in these fields where their study is unavoidable. There are also subjects that we have not touched on at all here, such as genetic drift and lottery models.

CHAPTER 4

Modelling structured populations

4.1 Modelling complex life cycles

The population models in the previous chapters have assumed that all the individuals are the same age or at the same stage in their life cycle. Here we will introduce models which can take account of individuals of different age, stage, or size. In particular we will use matrices to summarize the structure and parameters of a population composed of organisms with complex life histories. There is only space here for a short treatment of what is a rich and fascinating area of ecological modelling (Caswell 2000a).

Individuals of long-lived species may have widely varying patterns of pre-reproductive and reproductive life (Fig. 4.1). It will be assumed that, although generations overlap, reproduction occurs at certain times of year and there-fore discrete time models are appropriate. From the perspective of population dynamics there are two important differences between long-lived organisms with overlapping generations and annual or short-lived organisms with sepa-rate generations. First, long-lived organisms may delay reproduction for 1 or more years and, second, they may survive after reproduction to reproduce again. In all of these cases the life history of an individual may be categorized according to its age (e.g. time of first reproduction), stage (e.g. adult or juve-nile) or size (e.g. only plants over a certain size can reproduce).

Imagine a species, the individuals of which breed once a year, starting at age 3 years and which live to a maximum of 5 years. The reproduction and survival of these organisms can be described by a set of first-order difference equations. These give either the survival of individuals of different age or the reproductive output of individuals aged 3–5 years. Assume that the age-specific fecundity and survival parameter values are density-independent and are constant from year to year. For example, survival from birth to age 1 is described as:

Number of individuals aged 1 (in year $t+1$) = number born (aged 0 in year t)
    × fraction surviving from age 0 to 1

This can be represented algebraically:

$$N_{1,t+1} = N_{0,t}s_{0,1} \tag{4.1}$$

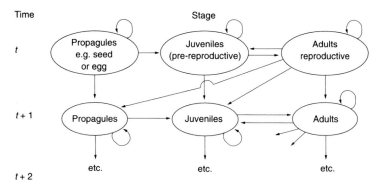

Fig. 4.1 Representations of life cycles of plant and animal species. Arrows show all possible transitions between stages, both within and between years.

In equation 4.1 the double subscript for the number of individuals (N) indicates the age class and the time (year). For the survival parameter (s) the double subscript describes the ages over which survival is considered. We can write similar equations describing the survival for the other age classes:

$$N_{2,t+1} = N_{1,t}s_{1,2} \tag{4.2}$$

$$N_{3,t+1} = N_{2,t}s_{2,3} \tag{4.3}$$

$$N_{4,t+1} = N_{3,t}s_{3,4} \tag{4.4}$$

$$N_{5,t+1} = N_{4,t}s_{4,5} \tag{4.5}$$

The fraction of individuals surviving from birth (age 0) to age 5 is therefore the multiple of the separate survival values from ages 0 to 1, 1 to 2 and so on; that is, $s_{0,1}s_{1,2}s_{2,3}s_{3,4}s_{4,5}$. We will assume that any individuals surviving to reproduce at age 5 then die. Therefore for any given value of N_0, N_5 could be predicted.

An equation is also required for the production of offspring (age 0 individuals in year t) by individuals aged 3–5 in the same year (t):

$$N_{0,t} = N_{3,t}f_3 + N_{4,t}f_4 + N_{5,t}f_5$$

f_3, f_4 and f_5 are age-specific fecundity parameters representing the average number of offspring per individual of that age in year t. Multiplying by $s_{0,1}$ gives an equation determining the number of offspring surviving to age 1 in year $t+1$ (see equation 4.1):

$$N_{1,t+1} = s_{0,1}(N_{3,t}f_3 + N_{4,t}f_4 + N_{5,t}f_5) \tag{4.6}$$

Equations 4.1–4.6 provide a complete description of the density-independent survival and fecundity of individuals in this age-structured population. We could explore by simulation the dynamics of this population, using these equations. Alternatively we can employ analytical techniques, in which case it is helpful to rewrite the equations in a different form, employing a matrix structure. As we do so, you might wish to consider whether you expect any fundamental differences in the dynamics of this population to the one described by equation 2.2.

Equations 4.2 to 4.6 can be represented as three matrices:

$$
\begin{pmatrix} N_1 \\ N_2 \\ N_3 \\ N_4 \\ N_5 \end{pmatrix} = \begin{pmatrix} 0 & 0 & s_{0,1}f_3 & s_{0,1}f_4 & s_{0,1}f_5 \\ s_{1,2} & 0 & 0 & 0 & 0 \\ 0 & s_{2,3} & 0 & 0 & 0 \\ 0 & 0 & s_{3,4} & 0 & 0 \\ 0 & 0 & 0 & s_{4,5} & 0 \end{pmatrix} \begin{pmatrix} N_1 \\ N_2 \\ N_3 \\ N_4 \\ N_5 \end{pmatrix}
\qquad (4.7)
$$

$$\mathbf{v}_{t+1} \qquad\qquad \mathbf{M} \qquad\qquad \mathbf{v}_t$$

Three matrices are required to summarize the five difference equations. There are two column matrices representing the number of individuals at ages 1–5 at times $t+1$ and t ($\mathbf{v}_{t+1}$ and $\mathbf{v}_t$ respectively). These column matrices are referred to as the population-structure vectors or age-distribution vectors. There is also one square matrix, $\mathbf{M}$, which gives all of the fecundity and survival values and is known as the population projection matrix. To check that equation 4.7 is equivalent to equations 4.2–4.6 you can multiply out the matrix and population-structure vector on the right-hand side of the equation. For readers unfamiliar with matrix multiplication, you begin by multiplying the five coefficients in the top row of the square matrix $\mathbf{M}$ by the corresponding population sizes in the column matrix $\mathbf{v}_t$ ($0 \times N_1$, $0 \times N_2$, $s_{0,1}f_3 \times N_3$ and so on) and add the resulting five multiplied pairs of values to give N_1 in $\mathbf{v}_{t+1}$. This process is then repeated with the next row, again multiplying by the corresponding values of N_1–N_5 in $\mathbf{v}_t$ and summing the five multiples. This process is repeated for all five rows of the matrix $\mathbf{M}$. Representation of age-structured populations in this manner was first described by Bernardelli (1941), Lewis (1942) and Leslie (1945, 1948).

Matrix equations such as equation 4.7, representing a set of difference equations, can be written in a general form to describe any age- or stage-structured population:

$$\mathbf{v}_{t+1} = \mathbf{M}\mathbf{v}_t \qquad (4.8)$$

where $\mathbf{v}_t$ and $\mathbf{v}_{t+1}$ are population vectors of the numbers of individuals at different ages (or sizes or stages) at t and $t+1$ respectively, and $\mathbf{M}$ is a square matrix in which the number of columns and rows is equal to the number of age classes. You will see the similarity of this to equation 2.2, $N_{t+1} = \lambda N_t$. This similarity is considered in the next section as we proceed with an analytical study of the dynamics of equation 4.8.

4.2 Determination of the eigenvalue and eigenvector

To proceed with the analytical investigation we will take a much simpler age-structured population and then discuss more complicated examples in the light of results from the simpler version. Consider a population of biennial plants (Fig. 4.2). The plant population has two age classes, which correspond to particular developmental stages. In the first year the plant forms rosettes following the germination of over-wintering seed. In the second year the surviving rosettes flower, set seed and then die. We will assume that the plant is a strict biennial: it always flowers in the second year (assuming it survives) and always dies after flowering. This model could also be described as a stage-structured population (Lefkovitch 1965; see Manly 1990 for an overview of matrix models of stage-structured populations) composed of rosettes and flowering plants. It is a coincidence in this case that each stage survives for one unit of time: in most cases this would not be true; for example, a tree species may spend many years at one defined stage. The dynamics of the population can be summarized with two first-order equations:

$$R_{t+1} = f s_{0,1} F_t \tag{4.9}$$

$$F_{t+1} = s_{1,2} R_t \tag{4.10}$$

where R is the number or density of rosette plants, F is the number of flowering plants, f is the average number of viable seed per flowering plant, $s_{0,1}$ represents the fraction of seed surviving between dispersal from the mother plant to rosette formation and $s_{1,2}$ describes the fraction of rosettes surviving until flowering.

In constructing such models it is often the case that stages such as seed are omitted. This will depend on the units of time chosen for the model and the census time. For example, we could have examined changes from spring to autumn and autumn to spring in which case seed may need to be included as a specified stage, or at least a seed/small rosette stage.

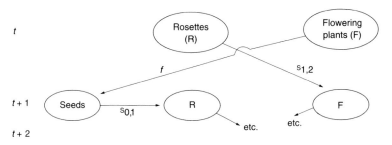

Fig. 4.2 Representation of the life cycle of a biennial plant species with fecundity (f) and survival at two different stages ($s_{0,1}$ and $s_{1,2}$). Seed as a separate stage is not included in this model.

As before, it is possible to write equations 4.9 and 4.10 in matrix notation (the algebraic shorthand for the matrices is indicated below them):

$$\begin{pmatrix} R \\ F \end{pmatrix} = \begin{pmatrix} 0 & fs_{0,1} \\ s_{1,2} & 0 \end{pmatrix}\begin{pmatrix} R \\ F \end{pmatrix} \tag{4.11}$$

$\quad\mathbf{v}_{t+1} \qquad\quad \mathbf{M} \qquad\quad \mathbf{v}_t$

We will now describe a mathematical analysis which will reveal two important results. First, it will provide the ratio of R to F, the composition or structure of the population. Second, it will give the finite rate of change of the biennial population, which will be seen to be equivalent to the finite rate of change (λ) in equation 2.2. Therefore this analysis makes the important assumption about the square matrix, $\mathbf{M}$, that it can be replaced by a single value (λ) and therefore that $\mathbf{Mv}_t = \lambda\mathbf{v}_t$. If this is true then the matrix equation 4.11 can be written as the density-independent equation 2.1, except now that $\mathbf{v}_{t+1}$ and $\mathbf{v}_t$ are population vectors rather than single numbers:

$$\mathbf{v}_{t+1} = \lambda\mathbf{v}_t \tag{4.12}$$

You should note that in multiplying the vector, $\mathbf{v}_t$, by λ, that all elements of the matrix are multiplied by λ. (λ is a scalar.) Equating the right-hand side of equations 4.11 and 4.12 – values at time t – we have:

$$\begin{pmatrix} 0 & fs_{0,1} \\ s_{1,2} & 0 \end{pmatrix}\begin{pmatrix} R \\ F \end{pmatrix} = \lambda\begin{pmatrix} R \\ f \end{pmatrix} \tag{4.13}$$

$\quad\quad \mathbf{M} \qquad\quad \mathbf{v}_t \qquad\quad \mathbf{v}_t$

It is helpful to have the right-hand side of equation 4.13 in a matrix form similar to the left-hand side. To do this we employ the identity matrix, $\mathbf{I}$. Multiplying any matrix by the identity matrix leaves the matrix unchanged (therefore $\mathbf{M} \cdot \mathbf{I} = \mathbf{M}$ on the left-hand side):

$$\begin{pmatrix} 0 & fs_{0,1} \\ s_{1,2} & 0 \end{pmatrix}\begin{pmatrix} R \\ F \end{pmatrix} = \lambda\begin{pmatrix} 1 & 0 \\ 0 & 1 \end{pmatrix}\begin{pmatrix} R \\ F \end{pmatrix}$$

$\quad\quad \mathbf{M} \qquad\quad \mathbf{v}_t \qquad\quad \lambda\mathbf{I} \quad\ \mathbf{v}_t$

Now multiply the identity matrix $\mathbf{I}$ by the scalar λ:

$$\begin{pmatrix} 0 & fs_{0,1} \\ s_{1,2} & 0 \end{pmatrix}\begin{pmatrix} R \\ F \end{pmatrix} = \begin{pmatrix} \lambda & 0 \\ 0 & \lambda \end{pmatrix}\begin{pmatrix} R \\ F \end{pmatrix} \tag{4.14}$$

$\quad\quad \mathbf{M} \qquad\quad \mathbf{v}_t \qquad\quad \lambda\mathbf{I} \quad\ \mathbf{v}_t$

We can now find a value for λ. Subtract the right from the left-hand side of equation 4.14:

$$\begin{pmatrix} 0 & fs_{0,1} \\ s_{1,2} & 0 \end{pmatrix}\begin{pmatrix} R \\ F \end{pmatrix} - \begin{pmatrix} \lambda & 0 \\ 0 & \lambda \end{pmatrix}\begin{pmatrix} R \\ F \end{pmatrix} = \begin{pmatrix} 0 \\ 0 \end{pmatrix}$$
$$\quad \mathbf{M} \qquad \mathbf{v}_t \qquad \lambda\mathbf{I} \quad \mathbf{v}_t$$

The left-hand side can be simplified by taking out the common vector ($\mathbf{v}_t$) and subtracting the two square matrices:

$$\begin{pmatrix} 0-\lambda & fs_{0,1}-0 \\ s_{1,2}-0 & 0-\lambda \end{pmatrix}\begin{pmatrix} R \\ F \end{pmatrix} = \begin{pmatrix} 0 \\ 0 \end{pmatrix}$$
$$\quad\; \mathbf{M} - \lambda\mathbf{I} \qquad\quad \mathbf{v}_t$$

to give:

$$\begin{pmatrix} -\lambda & fs_{0,1} \\ s_{1,2} & -\lambda \end{pmatrix}\begin{pmatrix} R \\ F \end{pmatrix} = \begin{pmatrix} 0 \\ 0 \end{pmatrix} \qquad (4.15)$$
$$\quad \mathbf{M} - \lambda\mathbf{I} \quad \mathbf{v}_t$$

If the matrix $\mathbf{M} - \lambda\mathbf{I}$ in equation 4.15 has an inverse then we could multiply both sides of the equation by the inverse matrix:

$$\begin{pmatrix} -\lambda & fs_{0,1} \\ s_{1,2} & -\lambda \end{pmatrix}\begin{pmatrix} a & b \\ c & d \end{pmatrix}\begin{pmatrix} R \\ F \end{pmatrix} = \begin{pmatrix} 0 \\ 0 \end{pmatrix}\begin{pmatrix} a & b \\ c & d \end{pmatrix}$$
$$\mathbf{M} - \lambda\mathbf{I} \quad \text{Inverse } \mathbf{v}_t \qquad \text{Inverse}$$
$$\qquad\;\; \text{of } \mathbf{M} - \lambda\mathbf{I} \qquad\quad \text{of } \mathbf{M} - \lambda\mathbf{I}$$

Multiplying the square matrix $\mathbf{M} - \lambda\mathbf{I}$ by its inverse on the left-hand side would give the identity matrix, $\mathbf{I}$ (by definition), whereas the right-hand side would reduce to 0:

$$\begin{pmatrix} 1 & 0 \\ 0 & 1 \end{pmatrix}\begin{pmatrix} R \\ F \end{pmatrix} = \begin{pmatrix} 0 \\ 0 \end{pmatrix}$$

$$\begin{pmatrix} R \\ F \end{pmatrix} = \begin{pmatrix} 0 \\ 0 \end{pmatrix}$$

This is unhelpful as we are left with the trivial solution that R and F are equal to 0. To overcome this problem we need to assume that the matrix $\mathbf{M} - \lambda\mathbf{I}$ does *not* have an inverse. This is true if the *determinant of the matrix is equal to 0*. This assumption can then be used to find a value for λ:

$$\begin{vmatrix} -\lambda & fs_{0,1} \\ s_{1,2} & -\lambda \end{vmatrix} = 0 \qquad (4.16)$$

The determinant in equation 4.16 is referred to as the *characteristic determinant*. The whole equation 4.16 is called the *characteristic equation*. We can now evaluate the characteristic determinant and therefore solve the characteristic equation:

$(-\lambda \cdot -\lambda) - f s_{0,1} s_{1,2} = 0$

$\lambda^2 = f s_{0,1} s_{1,2}$ (4.17)

We are now left with a quadratic equation (4.17). Initially this poses a problem because a quadratic equation has two solutions (or roots); in other words, λ can have two values. But earlier we had assumed that the square matrix **M** could be replaced by a single value, λ. Effectively this becomes true as the larges of the two λ values, referred to as the *dominant root*, has most influence on the dynamics. Note that the dominant root may be complex or negative. A negative dominant root is biologically meaningless in this application (but see Chapter 7) whereas complex roots are discussed in Chapter 7. In mathematics the values of λ are called the *eigenvalues* and the corresponding values of R and F are the *eigenvectors*. The eigenvalues may also be referred to as the latent roots or the characteristic values of the matrix, **M**. Similarly, the eigenvectors are known as the latent or characteristic vectors. (In passing it is worth noting that in finding values for R and F we have found solutions for the equations 4.9 and 4.10. Matrix methods have a wide application in the solving of simultaneous equations.) Finally, it may be helpful to know that equations 4.13–4.16 can be written in a general mathematical shorthand for any size of matrix **M** and vector **v** (as equation 4.8):

$\mathbf{M}\mathbf{v}_t = \lambda \mathbf{v}_t$

$\mathbf{M}\mathbf{v}_t - \lambda \mathbf{I}\mathbf{v}_t = 0$

$(\mathbf{M} - \lambda \mathbf{I})\mathbf{v}_t = 0$

The requirement for the non-trivial solution is that

$|\mathbf{M} - \lambda \mathbf{I}| = 0$

with values of λ being found by solution of the characteristic equation.

To reinforce all these theoretical points let us consider a specific example. If $f = 100$, $s_{0,1} = 0.1$ and $s_{1,2} = 0.5$ then from equation 4.17:

$\lambda^2 = 100 \times 0.1 \times 0.5$

$\lambda^2 = 5$

$\lambda = \pm\sqrt{5}$

$+\sqrt{5}$ is both the larger value (and therefore the dominant root) and the one which is ecologically meaningful. We can now use this value of λ to produce a prediction of the rate of increase of R and F (based on equation 4.12):

$\mathbf{v}_{t+1} = \sqrt{5}\mathbf{v}_t$

or

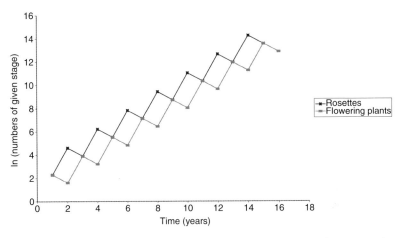

Fig. 4.3 Simulation of population dynamics of rosettes and flowering plants (equations 4.9 and 4.10) with values of $f = 100$, $s_{0,1} = 0.1$ and $s_{1,2} = 0.5$.

$$\begin{pmatrix} R \\ F \end{pmatrix} = \sqrt{5} \begin{pmatrix} R \\ F \end{pmatrix}$$

$\mathbf{V}_{t+1}$ $\mathbf{V}_t$

It is important to note that the model predicts that both R and F increase at the same rate of $\sqrt{5}$, and therefore predicts that they maintain the same ratio of R to F over time; that is, that they maintain a stable age structure. A quirk of this model is that it produces oscillations from year to year (Fig. 4.3). The yearly increase by $\sqrt{5}$ (λ) therefore needs to be viewed over a 2-year period; for example, from years 4 to 6 the rosette numbers increase 5-fold from 500 to 2500, equivalent to two yearly increases ($\sqrt{5} \times \sqrt{5}$).

We can quantify the eigenvector and therefore determine the ratio of R to F as follows, using equation 4.13:

$$\begin{pmatrix} 0 & fs_{0,1} \\ s_{1,2} & 0 \end{pmatrix} \begin{pmatrix} R \\ F \end{pmatrix} = \sqrt{5} \begin{pmatrix} R \\ F \end{pmatrix}$$

Using the given values for f, $s_{0,1}$ and $s_{1,2}$ and multiplying out the left- and right-hand sides:

$$\begin{pmatrix} 10F \\ 0.5R \end{pmatrix} = \begin{pmatrix} \sqrt{5}R \\ \sqrt{5}F \end{pmatrix}$$

In effect we now have two equations: $10F = \sqrt{5}R$ and $0.5R = \sqrt{5}F$. These two equations are equivalent because rearrangement of either produces $R = 2\sqrt{5}F$.

We have now achieved both parts of the analysis described at the beginning of this section: we have found a value for λ, the finite rate of change, by determining the eigenvalue of the matrix and we have calculated the ratio of R to F by quantifying the eigenvector.

These techniques can be applied to more complex examples in which there are more than two ages, stages or sizes of organisms. The number of eigenvalues is equivalent to the number of rows or columns and therefore the number of ages, stages or sizes in the projection matrix **M**. Although the determination of eigenvalues becomes more difficult as the matrix increases in size, the principle continues to hold that it is the dominant eigenvalue that is important. However large the projection matrix is, it can always be reduced to the dominant eigenvalue to describe the dynamics of the component stages of the population. Furthermore, the assumption of a stable age structure continues, given by the values in the eigenvector. Although we have focused on an age-structured population, it should be noted that many of the details of construction and results of the model are also relevant to stage- or size-structured populations.

The modelling of structured populations can be progressed by investigating the contributions of the various survival and fecundity values to the overall rate of change summarized by the eigenvalue (λ). These analyses have applications in harvesting and conservation of populations (Caswell 2000b). Sensitivity and elasticity are two related methods for determining contributions to the change in λ. Sensitivity quantifies the absolute changes in λ while elasticity quantifies relative changes in λ in response to proportional changes in elements of the projection matrix (de Kroon et al. 2000). Because the elasticity values sum to 1 the different components of the projection or transition matrix, such as the fecundity values, can be contrasted to show their importance to λ. This property has been used in comparative studies of life history across different taxa (e.g. Franco & Silvertown 2004).

4.3 Stochastic matrix models and succession

In Chapter 3 we saw how deterministic models of the form $N_{t+1} = \lambda N_t$ introduced in Chapter 2 can be developed by incorporating stochastic processes. Similarly, the deterministic matrix models outlined above have a stochastic counterpart (Fieberg & Ellner 2001) in which the various components of the matrix fluctuate in response to environmental change. These fluctuations are usually assumed to be like those of the random-walk example in Chapter 3; that is, independent and drawn from the same probability distribution. This type of process is referred to as a Markov process or Markov chain. The precise probability distribution used may vary within a given matrix or the probability distribution may be the same but the size of fluctuation may vary. There is also the possibility of including a deterministic signal. Fieberg and

Table 4.1 Fifty-year tree-by-tree transition matrix for grey birch, blackgum, red maple and beech. Each value is a transition probability.

Now	50 years hence			
	Grey birch (GB)	Blackgum (BG)	Red maple (RM)	Beech (B)
Grey birch	0.05	0.36	0.50	0.09
Blackgum	0.01	0.57	0.25	0.17
Red maple	0	0.14	0.55	0.31
Beech	0	0.01	0.03	0.96

Ellner (2001) discuss the ways in which stochastic matrix models can be used to estimate population extinction parameters.

Succession is the directional change in plant and animal species over time in a particular area. Mathematical models of this phenomenon have represented it as a Markov chain (Horn 1975, 1981). This involves determining the probability that a given plant (or other species or suite of species) will be replaced in a specified time by another individual(s) of the same or different species. Under Markov chain assumptions these replacement probabilities do not change with time. At each point in time, the relative abundances of species are multiplied by the transition probabilities to generate new relative abundances. This is iterated over a given number of time intervals. For example, Horn (1975, 1981) gave the values for 50 year tree-by-tree replacement between four species (Table 4.1). The model can be represented in matrix form:

$$\begin{pmatrix} GB \\ BG \\ RM \\ B \end{pmatrix} = (\text{transition probabilities}) \begin{pmatrix} GB \\ BG \\ RM \\ B \end{pmatrix}$$

$\mathbf{v}_{t+1}$ Transition probability
        matrix $\mathbf{v}_t$

These models predict a stationary end point; that is, that there will be a fixed ratio of grey birch to blackgum to red maple to beech. This is analogous to the result of a stable age structure in a population model. Interactions over different periods of time and the end point of the Horn example are given in Table 4.2. The predicted end-point is compared with the observed composition in old growth forest.

Other studies have looked at successional transitions between woodland and other types of vegetation. For example, Callaway and Davis (1993) used aerial photographs to measure transition rates between grassland, coastal sage scrub, chaparral and oak woodland and their relationship to burning

Table 4.2 Predicted composition of a succession at different time points.

Age of forest (years) . . .	0	50	100	150	200	End point	Observed very old forest
Grey birch	100	5	1	0	0	0	0
Blackgum	0	36	29	23	18	5	3
Red maple	0	50	39	30	24	9	4
Beech	0	9	31	47	58	86	93

Table 4.3 The percentage of vegetation type from aerial photographs in 1947 and 1989 in central coastal California.

Year	Vegetation (%)			
	Grassland	Coastal sage	Chaparral	Oak woodland
1947	21.5	26.4	28	24.1
1989	23.3	25.9	24	26.8

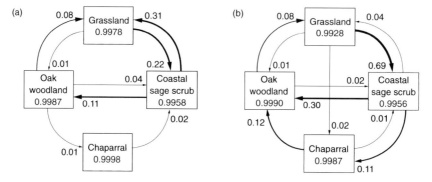

Fig. 4.4 Annual transition rates among plant communities in (a) burned plots ($n = 53$) and (b) unburned plots ($n = 78$) as determined from changes in vegetation between 1947 and 1989 shown on aerial photographs. The numbers in the boxes represent the probabilities that a given community will stay the same (from year to year) whereas the numbers on the arrows estimate the probability that a community will change in the indicated direction.

and grazing in Gaviota State Park in central coastal California, USA, between 1947 and 1989. The percentages of vegetation (community) types in 1947 and 1989 are given in Table 4.3 based on 0.25 ha plots sampled from aerial photographs.

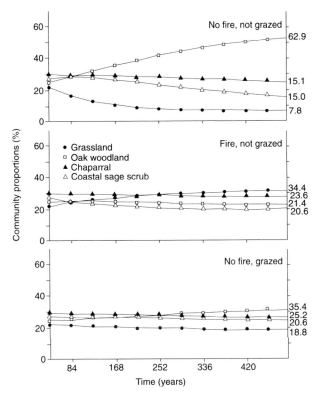

Fig. 4.5 Markov chain model predictions of future change in proportions of plant communities. Final community proportions at the end point (defined as <0.1% change over 42 years) are presented at the right, opposite each curve.

Although the overall percentage cover was very similar there was considerable flux between the years within plots. Transitions between vegetation type occurred in 71 out of 220 plots (32%). The transition probabilities were determined using these data (Fig. 4.4).

The current state could then be repeatedly multiplied by the four transitions (including no change) in the 42 year period to generate a Markov chain of predicted change in vegetation under particular environmental conditions. The predictions for three combinations of burning and grazing are given in Fig. 4.5.

Markov models are important tools in understanding landscape change and management. Modellers are using these tools in conjunction with statistical methods to assess spatial heterogeneity and rapidly improving data sets to provide more accurate predictions of primary and secondary succession (for example, Pueyo & Begueria 2007).

CHAPTER 5

Regulation in temporal models

5.1 Importance of density dependence

In the previous chapters we made some key simplifications concerning the dynamics of populations and clades. In particular, we assumed that under fluctuating environmental conditions the numbers of populations or lineages may drift continually upwards or downwards (until extinction) over time and that under constant conditions they may increase or decrease geometrically. Such continual drifting and/or geometric change is unrealistic under most conditions. Real populations often seem to be limited in their size and to be relatively abundant or relatively rare. The same may be true of clades in terms of the number of lineages. Populations that persist over long periods of time are presumed to be regulated in some way. The mechanism underlying this regulation is referred to as density-dependent change in survival or fecundity or, more succinctly, density dependence. For example, as the density of organisms increases there is an increase in the fraction of individuals dying (Fig. 5.1); that is, mortality is no longer constant for a particular age or stage of organism but is determined by the density of organisms (recall that density may refer to numbers or biomass per unit area or volume). Density dependence can be driven by processes such as competition or predation. For example, as population density increases, resources may become depleted and intraspecific competition become increasingly important, or predators may preferentially select prey at higher density. It is these assumptions of factors altering with population density that underlie the regulation of populations. Most ecologists agree that only density dependence can regulate populations. Later we will consider the extent to which clades may also show such density-dependent processes.

There are many examples of density dependence in the ecological literature (Fig. 5.1). These may be derived from laboratory or field experiments in which the density of organisms is altered (Figs 5.1a and 5.1b) or from natural variation in density in the field (Figs 5.1c and 5.1d).

An alternative to the detection of density dependence by experimentation is to examine the population dynamics for evidence of density dependence. The change in numbers from one time period (e.g. a year) to the next can be expressed as N_{t+1}/N_t or $\log N_{t+1} - \log N_t$. This change is plotted against density

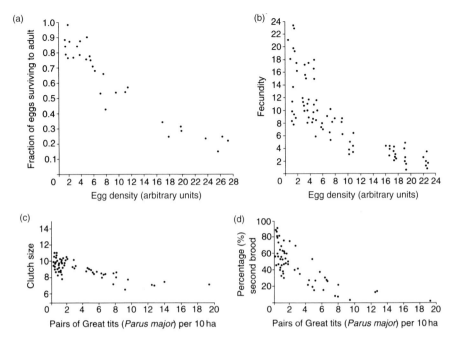

Fig. 5.1 Examples of survival and fecundity altering with population density in (a,b) *Drosophila melanogaster* (Prout & McChesney 1985) and (c, d) great tits (*Parus major*; data of Kluiver in Hutchinson 1978).

(N_t) to look for density dependence (Fig. 5.2). The null hypothesis is that there is no relationship between the change in population density and density itself. If density dependence is occurring then when density is low population size is likely to increase (N_{t+1}/N_t will be greater than 1) and when density is higher the population size is likely to decrease. Therefore we would expect a negative slope on a graph of change in population density against density under conditions of density dependence. The statistical significance of deviation from the null hypothesis of a gradient of 0 can be determined (although these analyses are problematic as we will see later). This method can also be applied within years to look for density-dependent survival or fecundity, both of which may contribute to a change in population numbers over time.

If we are to make our models more realistic then we must understand how density dependence affects population dynamics and incorporate it into these models.

5.2 Equations for modelling density dependence

The essence of density-dependent mechanisms for models of population dynamics is that, as the density increases, there is an alteration in the

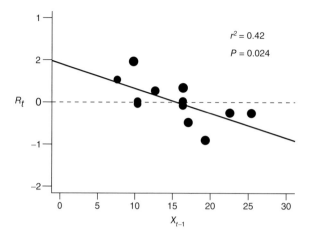

Fig. 5.2 Detection of density dependence by linear regression. In this example the population change in common sardine (*Strangomera bentincki*) from one year to the next ($R_t = \log X_t - \log X_{t-1}$) is regressed against its density in year $t-1$ (Pedraza-Garcia & Cubillos 2008). Note that the use of t and $t-1$ could be replaced by $t+1$ and t.

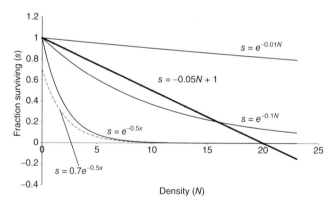

Fig. 5.3 Exponential and linear decline in fraction surviving (s) with density (N). Notice the change in shape with increasing values of a (see text for details). The intercept of the exponentially declining function can be easily altered (dashed line).

fecundity or fraction of individuals surviving and therefore a change in λ, the finite rate of population change. In open systems we can invoke density-dependent changes in migration, but these are not considered in this chapter. In Fig. 5.3 a set of lines have been drawn, illustrating some possible density-dependent functions (the vertical axis is labelled s to indicate the fraction surviving). In order to include these in population models we need to describe them in mathematical terms. For example, the linear decline in s with increasing density is expressed as the equation of a straight line:

$$s = c - mN \qquad\qquad\qquad (5.1)$$

where c is the intercept on the vertical (s) axis and m is the magnitude of the gradient of the line. c must lie between 0 and +1 as we are considering the fraction surviving. Equation 5.1 tells us that at the lowest (non-zero) density the value of s is close to maximum (c) and that with increasing density the fraction of individuals surviving decreases linearly according to the gradient m. The linear reduction in s with N assumes that there is a density, N_{max}, at which $s = 0$. In other words there is an upper (N_{max}) and lower boundary (0) of possible extinction. If a population goes above N_{max} it will become extinct. This upper boundary clearly creates some problems for modelling purposes. To overcome this we will consider a second mathematical function of exponential decline:

$$s = e^{-aN} \qquad\qquad\qquad (5.2)$$

The change in s with increasing values of N for three different values of a (0.5, 0.1 and 0.01) is shown in Fig. 5.3 (recall that equation 5.2 can be natural-log transformed to give $\ln(s) = -aN$, showing that the natural log of s declines linearly with N). The parameter a can be thought of as denoting the strength of density dependence. At any given value of N, the fraction surviving will decrease as a increases. When $N = 0$, s will be equal to 1, regardless of the value of a; that is, the model is designed so that at very low densities s tends towards a maximum value. Conversely, at very high densities, the value of s tends towards 0 but never reaches it (an example of an asymptote) unlike the linear density dependence. The function can be altered to $s = be^{-aN}$ to give a maximum value different from 1 at $N = 0$.

Some of the different curves in Fig. 5.2 can be considered to represent different types of intraspecific competition. An important distinction is between scramble and contest competition (Hassell 1975). In pure scramble competition, resources are divided equally among competing individuals. The consequence of this is that above a certain density the mean resource per individual is too low for survival, and therefore s plummets to zero. The upper boundary in the discrete logistic model (Section 5.4) could be interpreted as representing perfect scramble competition. The other extreme is contest competition, in which the superior competitors monopolize the resource. Consequently, a certain number of individuals always survive, even at high densities. In this case, s would approach zero at high densities, but never reach it, in agreement with the exponentially declining curve.

5.3 A density-dependent model of population dynamics

Let us return to the annual plant model from Chapter 2 and consider how density dependence may be incorporated into that density-*in*dependent model. Imagine that competition for space occurs between juvenile and adult plants: that there is intraspecific competition. The effect of competition on individual survival is expected to increase as plant density increases. In years of low plant density there will be relatively high survival but increasing density will result in reduced survival. We have seen that the simplest model for density dependence is a linear change represented by $s = -mN + c$ (equation 5.1). In the following example we will use d as the density of individuals subjected to density dependence. In the density-independent model we assumed that an average of 0.2 plants survived after germination up to seed set. In the new density-dependent model, 0.2 can be taken as the value of c; that is, when the effects of density are negligible. Therefore s is close to c at the lowest population densities and, with any increase in density, s declines linearly according to the gradient m. The linear reduction assumes that there is a density, d_{max}, at which $s = 0$, so that $m = 0.2/d_{max}$ (i.e. c/d_{max}). Therefore the linear density dependence equation can be written as $s = 0.2 - (0.2/d_{max})d$ or $s = 0.2(1 - d/d_{max})$. A new model, incorporating density dependence, now replaces the old density-independent model:

Number of germinating seed next year (N_{t+1}) = number of seed germinating this year (N_t) × fraction surviving to seed set ($0.2(1 - d/d_{max})$) × average number of seeds produced (100) × fraction surviving over winter (0.1)

If the fecundity and survival values are combined, as before, into the single value, λ ($=0.2 \times 100 \times 0.1$), we produce the equation:

$$N_{t+1} = N_t \lambda (1 - d/d_{max}) \tag{5.3}$$

In this case the interpretation of density dependence is that the fraction of germinating seed that survives is reduced by increasing density, which could be caused by intraspecific competition. In year t, d will be equal to $0.2N_t$. If $0.2/d_{max}$ is replaced by $1/K$ we obtain:

$$N_{t+1} = \lambda N_t (1 - N_t/K) \tag{5.4}$$

Equation 5.4, incorporating density dependence, is known as the discrete logistic equation and represents a strategic model for the population dynamics of annual species. K is the carrying capacity, defined as the maximum number of individuals a habitat can support. Equation 5.4 is sometimes written as $N_{t+1} = \lambda N_t (1 - \alpha N_t)$; that is, replacing $1/K$ by α. Berryman (1992) and Elliott (1994) review the use of the discrete logistic and similar equations whereas May et al. (1974) and May (1981) discuss the density-dependent terms.

Modelling of intraspecific competition has lead to a variety of equations incorporating density dependence. The model of Hassell (1975), described by the equation $N_{t+1} = \lambda N_t (1 + aN_t)^{-b}$, provides parameters a and b which can describe change from contest to scramble competition. a gives the threshold density at which density dependence occurs and b is the strength of the density dependence. This model, derived from earlier studies of insects (Morris 1959, Varley & Gradwell 1960), was related to models of fisheries ($N_{t+1} = \lambda N_t (1 + aN_t)^{-1}$ (Beverton & Holt 1957). In turn, the model of Hassell was developed by Watkinson (1980) to describe the population dynamics of annual plants:

$$N_{t+1} = \lambda N_t \Big/ \left((1 + aN_t)^b + w\lambda N_t \right)$$

where a and b are the parameters of the Hassell model, w is the degree of self thinning and λ is the finite rate of population change.

We will now consider some of the properties of equation 5.4 and compare them with the density-independent equations in Chapter 2. If we multiply out the right-hand side of equation 5.4 we see an important attribute of the density-dependent equation:

$$N_{t+1} = \lambda N_t - \left(\lambda N_t^2 / K \right) \tag{5.5}$$

Equation 5.5 clearly demonstrates that the discrete logistic is a nonlinear equation; in particular it is a quadratic equation, indicated by the presence of N_t^2. Its full title is a first-order nonlinear difference equation. It is first order because it relates values at time $t + 1$ to the previous time points (t). A second-order difference equation would relate values at time $t + 1$ to the previous two time points (t and $t - 1$). It is the non-linear component that gives this and similar equations (May 1976) some fascinating properties which we will now explore.

5.4 Exploration of the dynamics produced by the discrete logistic equation

We have two options in exploring the behaviour of equations such as the discrete logistic. First, given initial conditions, for example an initial number of germinating plants, and values for the parameters K and λ, we can generate a series of plant values as we did for the density-independent model. This is a simulation approach to the exploratory process: it will show us what the equation (model) can do but not necessarily tell us much about why it does it. If we want to know why, then we have to carry out some form of mathematical analysis, which is referred to as the analytical approach. Some analytical techniques are detailed after the simulations.

Values of N generated from simulations using equation 5.4 are displayed in Fig. 5.4. Starting with 10 germinating plants, Fig. 5.4a shows a flow diagram of the sequence of calculations in the simulation (such simulations can be written in widely available spreadsheet packages). This is an iterative process in which we generate a value for N_{t+1} and then use it as the new N_t and so on. You should check the first few iteration values given in Fig. 5.4b.

(a)

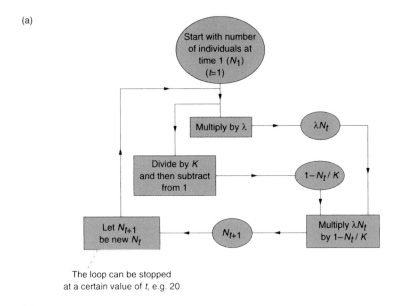

The loop can be stopped
at a certain value of t, e.g. 20

(b)

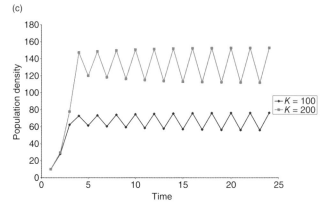

(c)

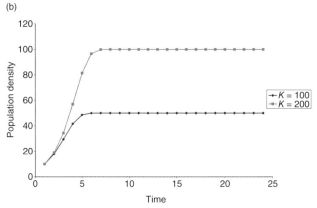

Fig. 5.4 (a) Flow diagram of the sequence of calculations showing how to generate successive values of N_t using the discrete logistic equation (equation 5.4). Change in density (N_t) with time generated from the discrete logistic equation with $K = 100$ and 200 and λ taking the values: (b) 2, (c) 3.1, (d) 3.5 and (e) 4. All graphs start with $N_1 = 10$.

Continued

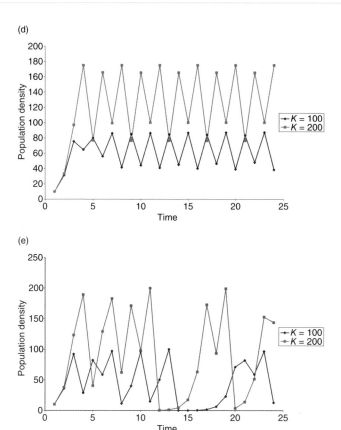

Fig. 5.4 *Continued*

The dynamics of the model population over time at different values of λ can be summarized as follows. At $\lambda = 2$ the population approaches an equilibrium value of $K/2$ at which it remains (Fig 5.4b); that is, this appears to be a stable equilibrium value. Recall from Chapter 2 that the equilibrium is defined as the population density to which or around which a population will move, whereas stability describes the tendency of a population to stay at or move towards or around the equilibrium. We have seen that density-independent dynamics can only produce a steady state if $\lambda = 1$. In contrast, density dependence allows an ecologically feasible stable equilibrium with different values of λ. At $\lambda = 3.1$ (Fig. 5.4c) the population oscillates between two densities; this is referred to as a two-point limit cycle. At $\lambda = 3.5$ (Fig. 5.4d) four-point limit cycles are produced whereas at $\lambda = 4$ (Fig. 5.4e) the initially regular cycles break up, so that the population fluctuates, apparently unpredictably, between a series of densities. This is referred to as chaotic dynamics: the mathematical definition of chaos and its importance in ecology is considered below. In this equation the values of K do not affect the dynamics and only contribute to the size of the equilibrium.

To be certain of the stability of the equilibrium with $\lambda = 2$ we need to displace the population from the assumed equilibrium and check its return. This can be achieved by running the model from different initial conditions and would show that the equilibrium of 50 is indeed stable; in fact it is globally stable for all ecologically realistic values. 'Global stability' has to be qualified to accommodate a flaw in the model, which is that it will crash if values of N_t exceed K (because $1 - N_t/K$ becomes negative). The two-point limit cycle is also stable; for a given value of λ (within the range of values giving two-point cycles) the population will always settle out to fluctuate between the same two densities so that there are now two stable equilibriums. In contrast, the chaotic dynamics do not have this property. Here, the particular sequence of values is dependent upon initial conditions, although the size of the fluctuations will be determined by the values of λ and K.

The possibility of chaotic dynamics means that if 'random' or unpredictable dynamics are recorded this does not necessarily imply that the underlying mechanisms are random (stochastic). Some or all of the 'randomness' could be produced by predictable deterministic processes expressed as chaos. Thus, if population change is described by the discrete logistic equation each population size at $t + 1$ (N_{t+1}) is given by a particular value of N_t. We can see this clearly by plotting N_{t+1} against N_t (Fig. 5.5a) using the parameter values for λ and K of 4 and 100 (Fig. 5.4e). The chaotic system shows the mathematical relationship of the discrete logistic: a quadratic equation. A fit through the points gives (as expected) a perfect fit indicated by the r^2 of 1. The coefficients of -0.04 and $+4$ agree with equation 5.5 ($-\lambda/K$ for N_t^2 and λ for N_t). This can be compared with a truly random sequence of values where N_{t+1} plotted against N_t is a cloud of points (Fig. 5.5b).

The challenge of detecting chaos in real population dynamics is therefore to distinguish it from random events. The first study to try to detect chaos in laboratory and field populations was by Hassell et al. (1976). They used the technique of assuming an underlying mathematical model (described by the equation $N_{t+1} = \lambda N_t(1 + aN_t)^{-b}$ discussed above) and determining the values of λ, a and b for different populations of insects. They were then able to compare these values with those known to produce limit cycles and chaos (Fig. 5.6). So Hassell and colleagues were testing whether the model that is fitted to the data has parameter values which would give chaos. The parameter values of b and λ for each species were superimposed on the regions of different dynamic behaviour predicted by the model; for example, stable equilibrium, limit cycles and chaos.

Only one species had values consistent with chaotic dynamics and one consistent with limit cycles. All the other populations were in the stable-equilibrium region. It is worth noting that the apparently chaotic population was a laboratory population of blowflies studied by Nicholson (1954). The

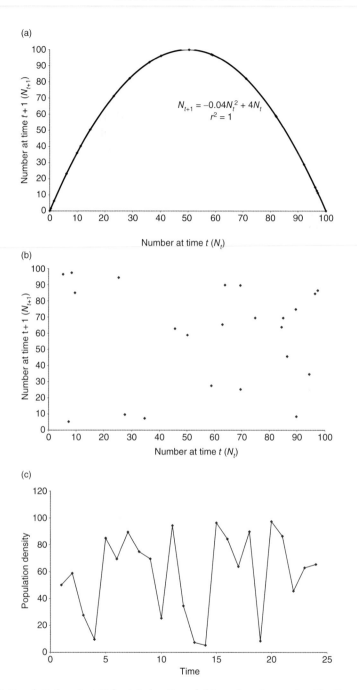

Fig. 5.5 N_{t+1} plotted against N_t for (a) chaotic and (b) random time series (the time series is shown in (c)). The chaotic time series is given in Fig. 5.4e.

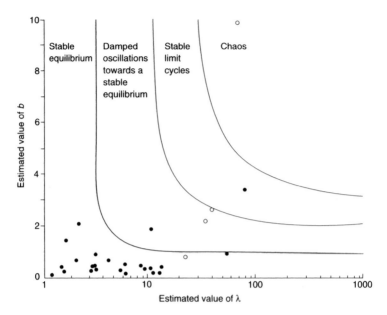

Fig. 5.6 Estimated values of λ and *b* (see text) for 28 populations of insect, overlaid on regions of different dynamic behaviour (open circles are data for laboratory populations, closed circles for field populations; Hassell et al. 1976).

debate on the importance of chaos in ecology rumbled on in the 1990s as new analytical techniques were explored. A study by Ellner and Turchin (1995) using three analyses suggested that Nicholson's blowflies were on the edge of chaos and no population (either in the laboratory or the field) was chaotic for all analyses. This might suggest that chaos is rare in ecological systems. However, extension to multi-species systems suggests that this is not the case (Chapter 7).

The possible presence of chaos in natural systems cautions us against assuming that all fluctuations are caused by stochastic events. The implication for population regulation is that density dependence is only stabilizing under certain conditions. If density dependence is coupled with high rates of population increase then chaos may result, which destabilizes the population.

We will now consider how to gain an analytical insight into the dynamic behaviour of the discrete logistic. First, the local equilibrium value for stable dynamics can be found. This is achieved by realizing that at equilibrium $N_{t+1} = N_t$; thus there is no change in the population density over time. If we replace N_{t+1} by N_t in equation 5.4:

$$N_t = N_t \lambda (1 - N_t/K) \tag{5.6}$$

we can then divide both sides by N_t to give $1 = \lambda(1 - N_t/K)$ and rename N_t as N^*, defined as the equilibrium value. This equation can be rearranged to make N^* the subject of the equation:

$1 = \lambda(1 - N^*/K)$

$1/\lambda = 1 - N^*/K$

$N^*/K = 1 - (1/\lambda)$

$N^* = K(1 - (1/\lambda))$ (5.7)

If we now substitute the values for K (100) and λ (2) we can see that $N^* = 100(1 - (1/2)) = 50$, which agrees with the result obtained in the simulation. The value of λ at which limit cycles begin can now be determined analytically. This is possible because it is known that limit cycles start when the gradient at equilibrium is equal to -1 (May & Oster 1976). For example, with the discrete logistic, we begin by differentiating equation 5.5 with respect to N:

$dN_{t+1}/dN_t = \lambda - (2\lambda N_t/K)$

$dN_{t+1}/dN_t = \lambda(1 - (2N_t/K))$

Set dN_{t+1}/dN_t equal to -1:

$-1 = \lambda(1 - (2N_t/K))$ (5.8)

Now solve this at equilibrium N^* by substituting $K(1 - (1/\lambda))$ from equation 5.7 for N_t in equation 5.8:

$-1 = \lambda(1 - (2K(1 - (1/\lambda))/K))$

Cancel the Ks to give:

$-1 = \lambda - 2\lambda + 2$

$-3 = -\lambda$

$\lambda = 3$

Therefore the single stable equilibrium ends and limit cycles begin at $\lambda = 3$, which agrees with the simulations in Fig. 5.4. Note again that the analytical method shows that the carrying capacity, K, is not relevant to the dynamics in this model.

A useful graphical method for analysis of first-order nonlinear difference equations is to plot N_{t+1} against N_t (a Ricker–Moran plot; more generally these are known as return maps and can be used with second- and higher-order processes). This was introduced above in the context of distinguishing between random and chaotic dynamics. These plots can be used to iterate the values of the time series as follows. Plot out the curve described by the equation, for example, at $\lambda = 3.1$ (Fig. 5.7) and then draw the line of $N_{t+1} = N_t$ (the line of unity). The point of intersection of the straight line and the curve tells us the value of N at which the equilibrium occurs. This is a graphical method of solving equation 5.6. To follow the dynamics on a Ricker–Moran

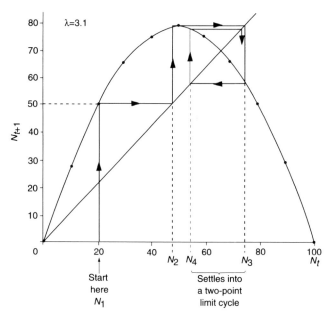

Fig. 5.7 Iterations of N_t on a Ricker–Moran plot using the discrete logistic equation with $\lambda = 3.1$ and $K = 100$.

plot we begin at an initial value of N_t, say 20. The next value of N (N_{t+1}) is read off the curve and, using the line of unity, N_{t+1} is changed to N_t and the process repeated. This eventually gives us the two-point limit cycles seen in Fig. 5.4c.

5.5 Re-evaluating the probability of extinction with density dependence

We will now introduce a stochastic element into the discrete logistic model in the same way that we did with the density-independent model (Chapter 3). λ is made to follow the probability density function as in Chapter 3 and the model is run for 20 generations with initial population sizes of one, five and 10. The results of 10 such simulations are illustrated in Table 5.1.

Comparison of the results in Table 5.1 with those in Table 3.1 shows that the probability of extinction has increased and mean persistence time has decreased. Indeed the probability of extinction is the maximum of 1 in each case. At first this seems counter-intuitive, as density dependence has been introduced into the model, which should have a stabilizing influence. However, the high probability of extinction is a result of the assumptions (or model flaw!) inherent in equation 5.4, in particular that at the maximum

Table 5.1 Simulation results for a density-dependent model (equation 5.4) with a probability density function (pdf) for λ as defined in Chapter 3.

Initial population size	Probability of extinction	Mean persistence time (generations)
1	1	4.3
5	1	6.8
10	1	10.3

density of K and above survival is zero. When close to K, the population may still leap above it in one generation, with the result that it becomes extinct in the next generation. So there is now an upper boundary as well as a lower boundary at which the population goes extinct. To remove the upper boundary at K and still retain the density dependence we need to replace the linear density dependence with a nonlinear function.

Instead of the linear density dependence we will use an exponentially declining function (Fig. 5.3) in which the fraction of individuals surviving (s) declines with density (N) according to equation 5.2 ($s = e^{-aN}$). This leads to a new equation representing the population dynamics of a species with non-linear density dependence:

$$N_{t+1} = \lambda N_t e^{-aNt} \tag{5.9}$$

Equation 5.9, like equation 5.4, is a first-order nonlinear difference equation and is known as the Ricker equation (Ricker 1954). This equation has been used extensively in fisheries studies to describe the relationship between recruitment (the number of offspring) and the size of the spawning stock (measured in terms of numbers or biomass). So these examples do not use a full life cycle (N_{t+1} as a function of N_t) but instead are considering density-dependent relationships between different stages or ages of a population. We will consider these functions in more detail in Chapter 6 when we discuss the concept of sustainable harvesting (Fig. 5.8 gives an example of a Ricker-type curve used to address concerns about a crash in the North Sea cod stocks; Cook et al. 1997).

Once again, simulations to determine the probability of extinction can be generated (Table 5.2) to compare with the two previous models (density-independent and linear density dependence). Both the probabilities of extinction and the mean persistence times are comparable with the density-independent model (Table 3.1). This is because the model has no upper extinction boundary (unlike the first density-dependent model) but has the same lower extinction boundaries as both previous models. Holmes et al. (2007) describe methods for statistical modelling of population extinction which address issues of density dependence, age structure and species interactions.

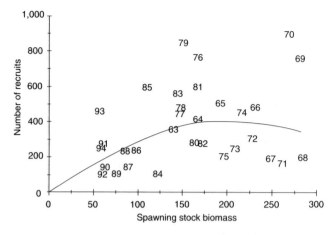

Fig. 5.8 Example of a Ricker-type curve used to address concerns about a crash in the North Sea cod stocks (Cook et al. 1997). The fitted curve is in fact a Shepherd stock-recruitment function ($R = aS/(1 + (S/b)^c)$) which produces a similar shape to the Ricker curve but in this case had a slightly better fit ($r^2 = 0.255$, r^2 for Ricker curve = 0.234). Numbers indicate years.

Table 5.2 Results from 10 simulations of a density-dependent model (equation 5.9) with $a = 0.001$ and a pdf for λ as defined in Fig. 3.7.

Initial population size	Probability of extinction	Mean persistence time (generations)
2	0.8	8.8
5	0.6	12.3
10	0.4	15.9

5.6 Estimation of parameters from field data and incorporation of density dependence at different stages in the life cycle

So far we have guessed the parameter values for the density dependence in our model, whereas λ has either been assumed to be constant or drawn from a hypothetical distribution. How can we use field data to estimate the parameter values of the model and therefore make them more realistic? It is possible to set up field or microcosm experiments in which population densities and habitat variables (such as food resources) are manipulated so that density-dependence parameters can be assessed. A second complementary method of estimating density dependence that can also be used to estimate values for λ is to use time-series data; for example, a series of annual censuses, preferably from several sites. Consider the density-dependent model described by the Ricker equation (5.9). A convenient way of estimating the two parameter values (λ and a) is to use linear regression. To start, divide both sides by N_t and take the natural logarithms:

$$\ln(N_{t+1}/N_t) = \ln(\lambda) - aN_t \qquad\qquad (5.10)$$

Now using linear regression of $\ln(N_{t+1}/N_t)$ against N_t we can estimate $\ln(\lambda)$ from the intercept and a from the gradient. With these estimated parameter values we can explore the dynamics of a given population under this model using either simulation or analytical techniques.

Woiwod and Hanski (1992) used regression of the Ricker equation as one of several measures of density dependence for 94 species of aphid and 263 species of moth collected in the Rothampsted insect survey. This comprehensive analysis showed that detection of density dependence increased with census duration to between 60 and 80% with the Ricker equation being one of two methods to give the highest detection rate (Fig. 5.9).

Although the regression method for detecting density dependence is straightforward we need to be cautious as this method can also detect density dependence from a time series composed of random values. Therefore it may detect density dependence when it is not there (Dennis & Taper 1994, Gillman & Dodd 2000).

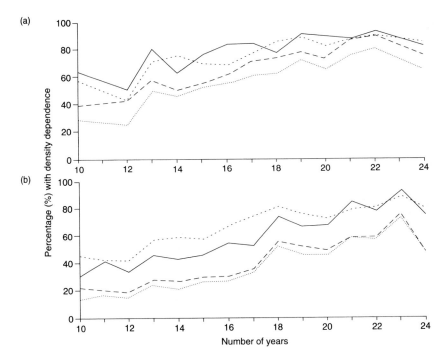

Fig. 5.9 Percentage of time series with significant ($P < 0.05$) density dependence detected using three methods in (a) aphids and (b) moths. Methods used were Bulmer's (———), Ricker's (· · · · · ·) and Pollard et al.'s (– – –). The methods are also shown combined (⎯⎯⎯⎯).

Density dependence may operate at one or more points in the life cycle, which is especially relevant when dealing with species with more complex life cycles. For example, plant species may experience density dependence at the seed, seedling and reproductive stages. Density dependence can operate among seeds if there is a limited number of microsites for germination and only one seed can germinate per microsite. If this is the case, as the density of seeds increases there will be a decreasing fraction that can germinate. Similarly, at the other end of the life cycle, if plants are insect pollinated and there is a fixed number of pollinators, higher densities of flowers may result in a decreased fraction of insect visits. In this case, there may be the opposite effect of enhanced visits to higher-density clumps, although the overall visitation decline will occur at some point of overall flower density.

Some studies have actively sought to investigate density dependence at different stages (Gillman et al. 1993). An example of the inclusion of density dependence in a stage-structured model is that of de Kroon et al. (1987). They investigated the effects of mowing as a management regime on the perennial rosette-forming *Hypochaeris radicata*. Density dependence was incorporated at both germination and seedling establishment, which were in turn functions of gaps in the vegetation. Their stage-structured model had four stages (seeds and three stages of rosette, the latter two of which were further divided into flowering and non-flowering; Fig. 5.10a). The time steps were between seasons rather than years. The authors also argued that a sigmoidal (s-shaped) rather than negative exponential form of density dependence was most appropriate (see Gillman & Crawley 1990 for an example of how tweaking a sigmoidal density-dependent function can lead to the different outcomes of stability, limit cycles and chaos). The results of their model under different mowing frequencies are shown in Fig. 5.10b. Increasing mowing frequency produced more gaps in the vegetation and higher germination, seedling establishment/survival and rosette survival. The net result was that population growth rates were predicted to be higher with increased mowing frequency. The highest mowing frequency produced damped oscillations in the model (Fig. 5.10b).

Given that density dependence can occur at different life-cycle stages and we know that it may cause different dynamical behaviours, it is of interest to know how different types of density dependence may interact. This issue has been explored by Buckley et al. (2001), who contrasted the behaviour of a single difference equation incorporating density dependence (the model of Hassell et al. 1976) and a second model which included a maximum of three density-dependent functions which affected probability of flowering, probability of survival to reproduction and fecundity per flowering plant. Both models performed well as a description of the dynamics while a systematic removal of the components of the second model showed the importance of the strong density dependence in fecundity. Other examples of modelling of annual plants and the importance of factors such as density

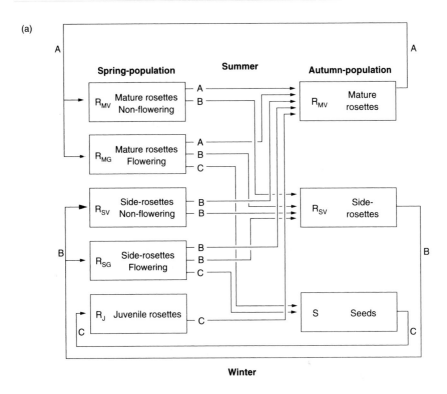

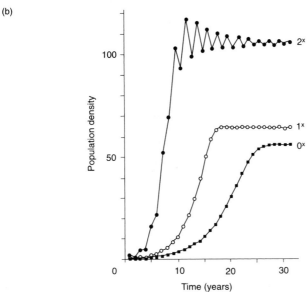

Fig. 5.10 (a) Life history of *Hypochaeris radicata* as used in the model of de Kroon et al. (1987). The three main life-history pathways are survival of adults (A), vegetative propagation (B) and sexual reproduction (C). (b) Simulated population growth of *H. radicata* with three mowing frequencies.

dependence, spatial dynamics and parameter uncertainty can be found in the review of Holst et al. (2007) and the discussion in Freckleton et al. (2008).

5.7 Density dependence in models of continuous populations

Both $N_{t+1} = \lambda N_t$ (equation 2.2) and $dN/dt = rN_t$ (equation 2.11) are density-independent equations because the growth parameters λ and r are unaffected by density. We can incorporate the ecological realism of density dependence into differential equations for population change in the same way that we did for the difference equation model in section 5.3 (equation 5.4). As there, the simplest of several possible forms of density dependence is a linear reduction in r with density. This is achieved by multiplying r by $(1 - N_t/K)$:

$$dN/dt = rN_t(1 - N_t/K) \tag{5.11}$$

Equation 5.11 is known as the logistic equation, in contrast to its discrete counterpart (equation 5.4). When N_t is close to 0, $1 - N_t/K$ is close to 1, so the value of r is relatively unaffected. r is steadily reduced as N increases until $N_t = K$ (carrying capacity). At this point there is no change in population size because $1 - N_t/K = 0$ and hence $dN/dt = 0$. If $N_t > K$ then $1 - N_t/K$ becomes negative, and therefore dN/dt is negative and population size decreases back towards $N_t = K$. Thus K is a stable equilibrium point. There is a smooth approach to the equilibrium value of K (Fig. 5.11a).

The logistic equation was first used by Verhulst (1838) to describe the growth of human populations and independently by Pearl and Reed (1920) to describe human population growth in the USA (Fig. 5.11b). Hutchinson (1978) and Kingsland (1985) provide fascinating insights into the history of this work. When we first considered the work of Pearl and Reed in Chapter 2 it was to estimate r. It was clear from the residuals of ln (population size) against time that the relationship was nonlinear. Pearl and Reed wanted to

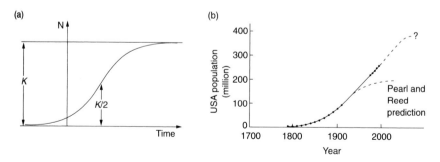

Fig. 5.11 (a) Shape of the logistic curve and relationship to parameters, reprinted from Pearl and Reed (1920). (b) Logistic growth of the human population of the USA from Pearl and Reed (1920) and subsequent data.

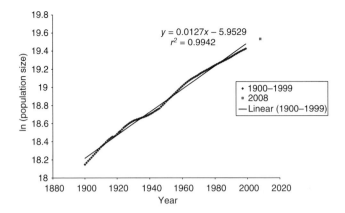

Fig. 5.12 Change in ln (population) size of USA during the twentieth century (www. census.gov/popest/archives/1990s/popclockest.txt).

find a general mathematical description of population growth which starts as exponential growth, but thereafter in their words 'must develop a point of inflection, and from that point on present a concave face to the x axis and finally become asymptotic, the asymptote representing the maximum number of people which can be supported on the given fixed area'. For this purpose they chose the logistic equation (although in a different form from that discussed here). Pearl and Reed fitted their equation to the population data in Fig. 2.13 to give values of $r = 0.0313$ and $K = 197\,000\,000$. The value of r is close to our linear estimate of 0.027 in Chapter 2. With the benefit of hindsight we know that, in April 2008, the current population of the USA exceeds Pearl and Reed's estimate of K by about 107 million (www.census.gov/ population/www/popclockus.html). The data set for the twentieth century summarized on the US census website shows an approximately linear increase in ln (population size) with time (Fig. 5.12). There is no evidence yet of an approach towards carrying capacity although there is a suggestion of slowing since about 1990. The linear fit gives a value of r of 0.0127 during the twentieth century.

The lack of agreement between Pearl and Reed's estimated maximum and the current population size for the USA is due to a variety of causes. For example, the logistic curve is symmetrical so that the population increase before the point of inflexion must equal the population increase after that point (Fig. 5.11a). This may be unrealistic for many population growth curves. Related to this point, Pearl and Reed's estimates were based on data prior to the point of inflexion. Human carrying capacity is also dependent on changing technology, so that predicted values of carrying capacity based on agricultural productivity and health care in the nineteenth century would inevitably be much lower following the green, biotechnology and medical revolutions of the twentieth century. Pearl and Reed, however, believed that their model provided a simple and useful description of the mechanisms underlying population growth (human or otherwise) and poetically invoked processes of density dependence and migration, as shown below.

> In a new and thinly populated country the population already existing there, being impressed with the boundless opportunities, tends to reproduce freely, to urge friends to come from older countries and by the example of their well-being, actual or potential, to induce strangers to immigrate. As the population becomes more dense and passes into a phase where the still unutilized potentialities of subsistence, measured in terms of population, are measurably smaller than those which have already been utilized, all of these forces tending to increase the population will become reduced.
> Pearl and Reed (1920), p. 287

The logistic curve has also been fitted to the growth of nonhuman populations, such as bacteria or yeast, with varying success. The best examples are cultures of algae, bacteria, insects and yeast (Fig. 5.13) where the simple assumptions of the logistic equation are most appropriate. The examples in Fig. 5.13 illustrate the sensitivity of r and K to genotype and environmental conditions.

Finally, given the models for the USA population it is of interest to consider what is happening to the world population (Fig. 5.14). Here, the raw data suggest that population increase has been approximately linear since the 1960s. This would suggest that the per-capita rate of growth is slowing. This is clear when we examine the ln (population size) (Fig. 5.15). Fitting a quadratic equation to the data since 1960 provides a very good fit ($r^2 = 0.9999$). Using the quadratic function we can predict the increase and point of maximum population size assuming the same growth pattern since 1960. This is shown for the raw data (Fig. 5.16). The maximum value of 9.172 billion is predicted to occur in the year 2062. The addition of approximately 2.7 billion more people is clearly going to place profound stresses on the resources of the planet.

5.8 Density dependence in diversification rate

The importance of density dependence in population dynamics is reflected in studies of cladogenesis. In Chapters 2 and 3 we saw how clades might be expected to grow exponentially. Just as it is unrealistic to expect populations to continue to grow exponentially so it is unreasonable to expect clades to continue to diversify exponentially. As niche space becomes filled we may expect diversification rates to slow down. In this case the density is a density of species, rather than of individuals in the case of population dynamics. Tests of this hypothesis took on a new impetus in the 1990s with the advent of molecular phylogenies. In a seminal paper, Nee et al. (1992) demonstrated

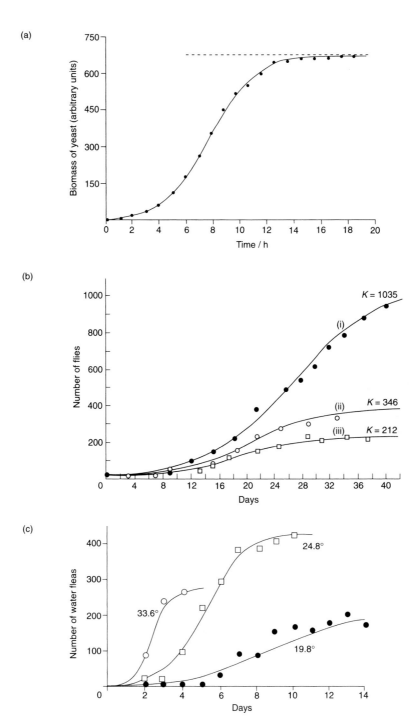

Fig. 5.13 Examples of applications of the logistic growth curve. (a) Growth of yeast populations in culture. From Allee et al. (1949) reprinted in Maynard Smith (1974). (b) Growth of *Drosophila melanogaster* populations: (i) wild type, (ii) heterozygous or homozygous individuals for five recessive mutations including vestigial wing and (iii) wild type in half volume of (i). (c) Growth of *Moina macrocarpa* populations at three different temperatures. (b) and (c) reprinted from Hutchinson (1978).

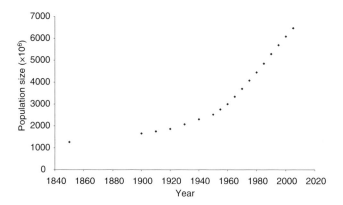

Fig. 5.14 Change in global human population size (raw numbers) from the mid-nineteenth to the early twenty-first century.

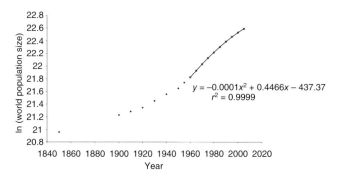

Fig. 5.15 Natural log of world human population size.

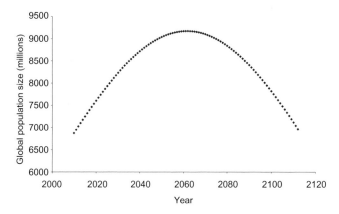

Fig. 5.16 Predicted human population size in the twenty-first to early twenty-second centuries based on the quadratic regression in Fig. 5.15.

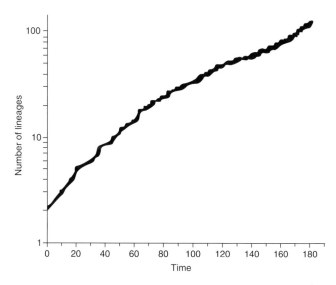

Fig. 5.17 Number of bird lineages against time. From Nee et al. (1992), Fig. 1.

a smooth reduction in the rate of diversification of birds (Fig. 5.17) using the molecular phylogeny of Sibley and Ahlquist (1990). The reduction of diversification rate was modelled by dividing the average diversification rate (p) by a function of the density of lineages (N) at that time; that is, p/N^a.

The analysis of Nee and colleagues has been supported by more recent work using improved phylogenies for a range of bird clades (Phillimore & Price 2008). This analysis took into account the possibility that, even with the same diversification rate, larger clades are expected to slow down more than small clades. Despite this, 57% of the large clades studied (more than 20 species) showed a significant slow down in diversification.

CHAPTER 6

Modelling interactions

I agree with him [Lotka] in his conclusion that these studies and these methods of research deserve to receive greater attention from scholars, and should give rise to important applications.
Vito Volterra (from Lotka 1927)

6.1 Overview of interactions

The dynamics of populations are affected by a variety of interactions with other populations (which are themselves dynamic). We begin by considering predator–prey interactions, encompassing all interactions between an organism and its natural enemies, specifically plant–herbivore, host–parasitoid (an insect that feeds in or on its host, leading to the host's death), herbivore–carnivore and host–pathogen interactions. Whereas each of these interactions has been modelled independently (and we will consider examples of these later) there are a set of results relevant to all predator–prey interactions which we will explore first. Foremost among these is the propensity of predators and their prey to cycle in abundance. The phenomenon is found in both invertebrates and vertebrates (Fig. 6.1).

Not all predators and prey show such cycles and some species do it in one part of their range but not in others. In the first part of this chapter we will use models to help us understand the dynamics of cycling in predators and prey and the variation within and between species. We will also consider some important applications of the dynamics of predator–prey interactions. These include the sustainable harvesting of animals and plants for food and the biological control of pest species; that is, the introduction of a natural enemy to control a pest species and control of human disease.

6.2 Cyclical dynamics arising from predator–prey interactions

6.2.1 Early continuous-time models

The origins of modelling of predator–prey dynamics are to be found in the independent work in New York of Lotka (1925) and in Rome of Volterra

(a)

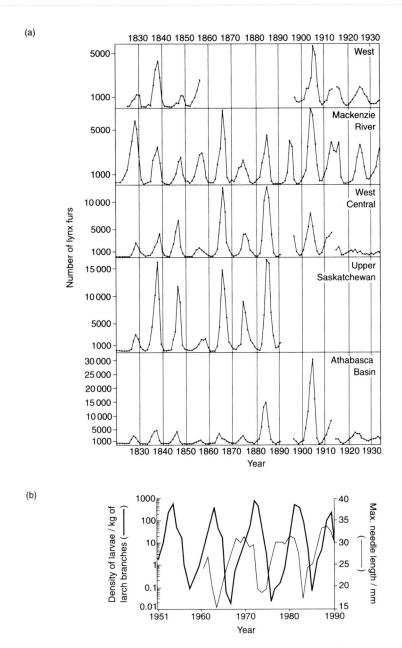

Fig. 6.1 Examples of cycling of abundance of predators and prey. (a) Cycles in the number of lynx (*Lynx canadensis*) fur returns of the Hudson's Bay Company, from 1821 to 1934, grouped into five regions. Note the different scales (Elton and Nicholson 1942). (b) Cycles of abundance of the monophagous larch bud-moth (*Zeiraphera diniana*) and the larch (variation in needle length). An 8–9 year cycle with a 10 000-fold change in larval density from peak to trough has occurred 16 times since 1850 in Switzerland (Baltensweiler 1993).

(1926, 1928), who also derived equations to describe competition. The independence of their work is illustrated in their communications to the journal *Nature* from which the opening quotation to this chapter is taken (Lotka 1927). The premise of their predator–prey models, framed in continuous time, was that of:

> ... two associated species, of which one, finding sufficient food in its environment, would multiply indefinitely when left to itself, while the other would perish for lack of nourishment if left alone; but the second feeds upon the first, and so the two species can coexist together.
> Volterra (1926)

Lotka and Volterra originally assumed that prey density (N) increased exponentially, quantified by r_1, in the absence of predators (see also equation 2.10):

$$\frac{dN}{dt} = r_1 N \tag{6.1}$$

This was made more realistic by Volterra by assuming that change in prey density was described by the logistic equation that we met in Chapter 2; that is, the prey population would move towards an equilibrium of K in the absence of predation:

$$\frac{dN}{dt} = r_1 N(1 - N/K) \tag{6.2}$$

In the presence of predators the rate of change of prey population size with time, dN/dt, is assumed to be reduced in proportion a to the density of predators (P) multiplied by the density of prey (N):

$$\frac{dN}{dt} = r_1 N(1 - N/K) - aPN \tag{6.3}$$

As we are modelling a dynamic system in which the predator population density may also fluctuate, we need to develop an equation for dP/dt. Lotka and Volterra assumed that, in the absence of prey, the predator population size would decline exponentially, quantified by r_2; that is, they assumed that the predator species specialized on one species of prey:

$$\frac{dP}{dt} = -r_2 P$$

In the presence of prey, this decline would be counteracted by an increase in predator density, again in proportion to the density of predators (P) multiplied by the density of prey (N):

$$\frac{dP}{dt} = -r_2 P + bPN \tag{6.4}$$

Equations 6.3 and 6.4 provide a system of two coupled first-order nonlinear differential equations. In section 6.2.2 we will consider a graphical technique for analysing the behaviour of coupled differential equations. This technique is very useful because systems of differential equations may arise in all the types of interaction considered in this chapter.

6.2.2 Phase plane analysis

To understand the dynamics produced by the Lotka–Volterra equations we will examine their behaviour in a phase plane where the densities of predator and prey at time t are plotted and linked to their rate of change at those densities. This method of analysis was developed by Rosenzweig and MacArthur (1963). A phase plane can be used to describe change in coupled differential equations.

Begin at time 1 with a value of P_1 for predators and N_1 for prey (Fig. 6.2). To these densities we attach a vector showing the change in P and N from that point. The vector is a combination of two orthogonal vectors (vectors at right angles): one representing the value of dP/dt and one representing the value of dN/dt. The two vectors are added together to give the resultant vector (Fig. 6.2).

It is not necessary to know the precise direction of change from any given combination of N and P. Indeed, the beauty of phase-plane analysis is that the dynamics of the system can be understood by sketching some of the vectors based on the signs; that is, positive or negative values of dP/dt and dN/dt. Assume the following parameter values for equations 6.3 and 6.4:

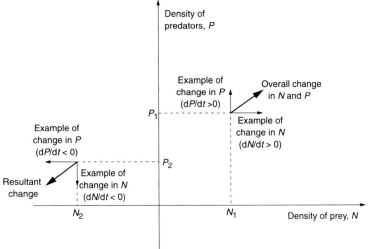

Fig. 6.2 Prey (N) and predator (P) densities and representation of associated dynamics in the phase plane.

$r_1 = 3$ and $r_2 = 2$ (prey > predator), $a = 0.1$, $b = 0.3$ and $K = 10$. Therefore $dN/dt = 3N - 3N^2/10 - 0.1PN$ and $dP/dt = -2P + 0.3PN$.

Initially, we need to determine when $dN/dt = 0$ and $dP/dt = 0$ (known as zero-growth isoclines). First, $dN/dt = 0$ when $3N - 3N^2/10 - 0.1PN = 0$. This is equivalent to $(3 - 3N/10 - 0.1P)N = 0$ so that either $N = 0$ (the trivial solution in which prey are absent) or $3 - 3N/10 - 0.1P = 0$. The latter can be rearranged to give $P = 30 - 3N$. This is a straight line equation that can be plotted on the phase plane (Fig. 6.3).

Similarly, $dP/dt = 0$ when $-2P + 0.3PN = 0$ or $P(-2 + 0.3N) = 0$. The solutions are either $P = 0$ (trivial solution) or $-2 + 0.3N = 0$; $N = 2/0.3 = 6.667$. This solution is plotted on the phase plane as a vertical line. The intersection of the two lines representing $dN/dt = 0$ and $dP/dt = 0$ is important, as this is where neither P nor N changes in density. The stability of this equilibrium point will be seen to be of great significance to the dynamics of the predator–prey system. The phase plane is divided up into four regions produced by the intersection of $dN/dt = 0$ and $dP/dt = 0$. Each region is characterized by a particular combination of positive or negative values of dN/dt and dP/dt (Fig. 6.4).

This is helpful as any point (N,P) in a given region will have a particular combination of vectors attached to it. Although the relative magnitudes of the vectors will depend on where the points are in the region, the overall direction of change in N and P (represented by the resultant vector) will always be the same. These directions are indicated as thicker arrows in Fig. 6.4.

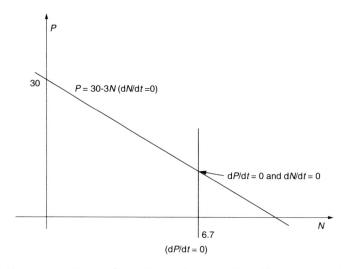

Fig. 6.3 Lines representing no change in prey density ($dN/dt = 0$) and predator density ($dP/dt = 0$) on a phase plane.

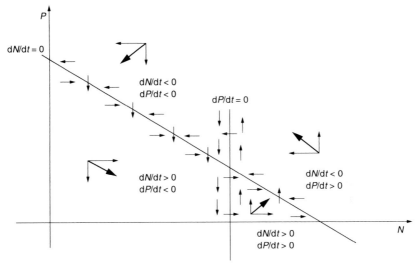

Fig. 6.4 Overall direction of change in prey and predator densities in four regions of the phase plane.

So, to find the directions of the vectors in any region we need to determine whether dN/dt and/or dP/dt are positive or negative. For dP/dt, if $N > 6.7$ then $-2 + 0.3N > 0$ and dP/dt is positive (and vice versa). Therefore to the left of the dP/dt zero-growth isocline (at $N = 6.7$) all the change in P is negative (arrows point down in Fig. 6.4), whereas to the right the arrows point up. When $dP/dt = 0$ there is no change in P, so there can only be change in N, as indicated by the horizontal arrows. The direction of the horizontal arrows is only known when the regions of dN/dt greater or less than zero have been determined.

Now consider the two regions either side of $dN/dt = 0$. $dN/dt < 0$ occurs above the line of $dN/dt = 0$. You can check this by substituting values for N and P; for example $P = 40$ and $N = 0$ in the inequality $3 - 3N/10 - 0.1P < 0$ gives $3 - 0 - 4 = -1$. Therefore above the line of $dN/dt = 0$ the horizontal arrows point to the left and below the line they point to the right. When $dN/dt = 0$ there is no change in N so there can only be change in P, indicated by the vertical arrows with the direction determined by whether they are to the left or right of $dP/dt = 0$.

Now, for any point (N,P) in the phase plane we know the direction of change. Taking any starting point on the phase plane it is possible to look at the dynamics of N and P as a trajectory across the phase plane. With the above parameter values, starting at any point away from the equilibrium will send the population spiralling into the equilibrium point. An example of a trajectory under these conditions is given in Fig. 6.5.

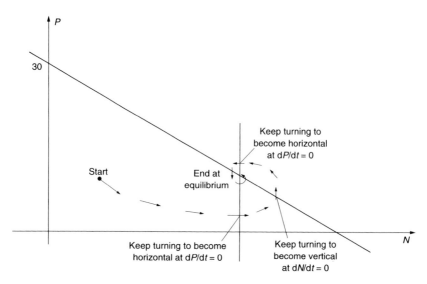

Fig. 6.5 An example of a plot of the trajectory of predator and prey dynamics on a phase plane.

If the $dP/dt = 0$ line is kept vertical and the slope of $dN/dt = 0$ is altered, what is the effect on dynamics and stability? In particular, what happens when $dN/dt = 0$ is horizontal, which is equivalent to removing the prey density dependence? In this case, starting at any point simply sends the trajectory on an elliptical path back to where it started (Fig. 6.6a). This is an example of neutral stability, for which the Lotka–Volterra model received much criticism. Neutral stability, akin to a frictionless pendulum, is referred to as a structurally unstable model (May 1973a), in which the amplitude of cycles is determined by initial conditions and the cycles persist with unchanging amplitude. This is in contrast to the limit cycles in Chapter 5 where the cycles fluctuate between particular densities regardless of the initial conditions; that is, the starting density. If the gradient of $dN/dt = 0$ is positive then the trajectory spirals out leading to unrealistically high values of predator and prey (Fig. 6.6b). When the gradient of dN/dt is negative, as in the above example, the result will be a stable equilibrium (Fig. 6.6c).

In fact, with the Lotka–Volterra equations, the only cycles produced are neutrally stable. Although Volterra considered the possibility of using equation 6.2 in his predator–prey model, noting that the fluctuations would be damped and the system would tend towards the stationary state, he did not pursue this line of enquiry. Most of his emphasis in predator–prey systems was on the possibility of cycles using equation 6.3 without prey regulation and equation 6.4. For structurally stable cycles we need a time delay of some description, which may be expressed either as a delay differential equation (Hutchinson 1948) or, as in the next section, by coupled difference equations.

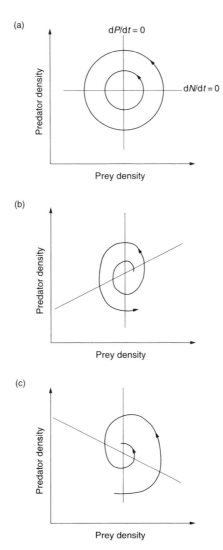

(a)

$dP/dt = 0$

Predator density

$dN/dt = 0$

Prey density

(b)

Predator density

Prey density

(c)

Predator density

Prey density

Fig. 6.6 Effect of changing the gradient of $dN/dt = 0$ on stability in Lotka–Volterra models. (a) Neutrally stable cycles of predators and prey (gradient of $dN/dt = 0$). (b) Densities of predators and prey spiral out (gradient of $dN/dt = 0$ is positive). (c) Densities of predators and prey spiral into a stable equilibrium (gradient of $dN/dt = 0$ is negative). Reprinted from Maynard Smith (1974).

6.2.3 Explaining cycles of abundance with models in discrete time

We will now consider a discrete version of the Lotka–Volterra model (May 1973b) and then generalize this to a new type of equation. By analogy with equations 6.3 and 6.4:

$$N_{t+1} = \lambda_N N_t(1 - N_t/K) - aP_tN_t \tag{6.5}$$

$$P_{t+1} = \lambda_P P_t + bP_tN_t \tag{6.6}$$

Therefore, in the absence of predation the prey are self-regulated according to the discrete logistic (equation 6.5, with the finite rate of prey population change given by λ_N), and in the absence of prey the predators decline at the rate of λ_P (if it is assigned a value less than 1). If the time subscript for the predator equation 6.6 is reduced on both sides by 1:

$$P_t = \lambda_P P_{t-1} + bP_{t-1}N_{t-1}$$

and the right-hand side substituted for P_t in equation 6.5:

$$N_{t+1} = \lambda_N N_t(1 - N_t/K) - a(\lambda_P P_{t-1} + bP_{t-1}N_{t-1})N_t \tag{6.7}$$

then we are left with a second-order nonlinear difference equation. The equation is second order because N_{t+1} is explained by terms which are two time steps (P_{t-1} and N_{t-1}) earlier, along with a term one time step earlier, N_t.

Therefore, a pair of coupled first-order difference equations is equivalent to a second-order difference equation. The dynamics produced by second-order nonlinear difference equations are very interesting and quite different from their first-order cousins. Second-order equations can produce cycles which are similar to those of predators and prey observed in the field. To test this type of model we need to know a little more about the mechanisms underpinning predator–prey interactions.

As an example consider the dynamics of the larch bud-moth (*Zeiraphera diniana*), which periodically defoliates its host tree, larch (*Larix decidua*), and the pine looper moth (*Bupalus piniaria*), which specializes on Scots pine (*Pinus sylvestris*). Detailed sampling of the larvae of the larch bud-moth has shown that the optimum area for survival and fecundity is the subalpine region of the Swiss Alps between 1700 and 2000 m (Baltensweiler 1993). In this region the larch bud-moth reaches carrying capacity within four or five generations and has cycle lengths of 8–9 years (Fig. 6.1). The clear cycles of *Zeiraphera* are in contrast to the rather irregular and sometimes absent cycles of *Bupalus* (Fig. 6.7).

Larch bud-moth cycles are an example of a herbivore population cycle explained by density-dependent changes in their food which may be modified by climate conditions. As larval densities increase, defoliation of the larch affects its physiology, reducing the quality and quantity (needle length) of the herbivore's food that, in turn, triggers the collapse of the herbivore populations in the following years. Alteration in the plant (food) quality includes decreased nitrogen content and greater fibre content. This explanation of cycles has been contested by some ecologists, who claim that interactions between herbivores and their natural enemies generate the cycles. Note that forest trees, as prey, may fluctuate or cycle in the abundance of biomass or components such as nitrogen, rather than numbers.

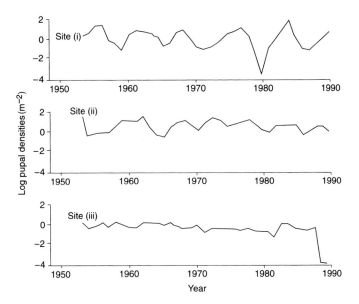

Fig. 6.7 Cycles of population abundance of pine looper (*Bupalus piniaria*) at three sites.

The larch bud moth example is a clear case of delayed density dependence in which the effects of density dependence are carried over to the next generation via an interaction with the host plant. Delayed density dependence, first described by Varley (1947), underpins the dynamics of many predator–prey systems and can be described by second-order nonlinear difference equations. Broekhuizen et al. (1993) identified a delayed density-dependent mechanism by which population cycles may be produced in the pine looper moth, detailed below.

> There is a strong negative correlation between annual growth increment of *Pinus sylvestris* in Tentsmuir forest and the pupal density of *Bupalus piniaria* in the previous two years (Straw 1991). This indicates that *B. piniaria* may have a substantial influence upon their host trees' physiologies. This may 'feed back' upon the *B. piniaria* population such that the one year's *B. piniaria* [population] influence the reproductive success of the next year's population.

To explore models of delayed density dependence we will consider a second-order nonlinear difference equation which can be parameterized from census data. This covers the work of Turchin (1990; see also Turchin & Taylor 1992) who used a second-order Ricker equation to investigate the likelihood of delayed density dependence among herbivorous forest insects, including *Zeiraphera* and *Bupalus*. We met the first-order version of the Ricker equation in Chapter 5 and will use the same regression technique to estimate parameters. The second-order version is:

$$N_{t+1} = N_t e^{(r + aN_t + bN_{t-1})} \tag{6.8}$$

The parameters a and b represent the strength of direct and delayed density dependence respectively. Note that the right-hand side of equation 6.8 is equivalent to $\lambda N_t e^{(aN_t + bN_{t-1})}$. If $b = 0$ then equation 6.8 reverts back to the first-order equation. For regression purposes we also assume an unexplained variance term in the model. To estimate the parameters a and b in equation 6.8 we divide by N_t and take natural logs:

$$\ln(N_{t+1}/N_t) = r + aN_t + bN_{t-1} \tag{6.9}$$

$\ln(N_{t+1}/N_t)$ can then be regressed against N_t and N_{t-1}. This was the method used by Turchin (1990), who showed significant delayed density dependence (i.e. values of b significantly different from 0) in 10 out of 14 data sets. In considering the significance of the parameters we should recall the possibility of overestimating the significance of a or b using this regression model. The values of a, b and r for *Bupalus* and *Zeiraphera* are given in Table 6.1.

The dynamics resulting from the estimated parameter values can be found by simulation using equation 6.8 (Fig. 6.8). You will see that the predicted dynamics for the two species are quite different, with *Bupalus* predicted to be stable with an equilibrium population size of about 150. *Zeiraphera*, in contrast, is predicted to cycle with periods of 6–7 years, slightly shorter than the 8–9 year cycles observed in the field. These results are in agreement with the clearly defined cycles of *Zeiraphera* and poorly defined cycles of *Bupalus*. Analysis of equation 6.9 therefore suggests a much stronger deterministic signal for cycle production in *Zeiraphera* compared with *Bupalus*.

Simulation of another second-order model in Broekhuizen et al. (1993) also suggests a stable equilibrium for *Bupalus* populations. The critical parameter determining cyclical behaviour in equation 6.8 is r. Indeed, whereas 10 out of 14 of Turchin's data sets showed delayed density dependence, only three have values of r high enough to produce cycles. Turchin and Taylor (1992) noted that focusing on direct density dependence rather than delayed density dependence leads to potentially misleading results. They gave the example of the analysis of Hassell et al. (1976), who estimated density dependence and then looked at the stability (or otherwise) of insect population

Table 6.1 Parameter values derived from regression of equation 6.9 (Turchin 1990). Both values of a and b are significantly different from 0. Variance (%) is the percentage of variance in the dependent variable explained by variance in the independent variable (recall that this is a measure of goodness of fit of the regression and written as r^2; not to be confused with the r in column 2, which is the intrinsic rate of change).

	r	a	b	Variance (%)
Bupalus	0.34	−0.0005	−0.0018	24.1
Zeiraphera	1.20	−0.0001	−0.02	51.1

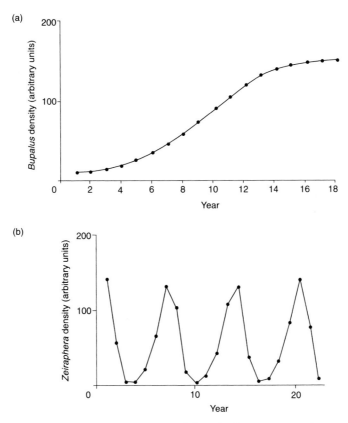

Fig. 6.8 Predicted dynamics of (a) *Bupalus* and (b) *Zeiraphera* using equation 6.8 and the parameter values in Table 6.1. The figure for *Bupalus* shows the population moving smoothly (and sigmoidally) from initial conditions of $N_1 = 10$ and $N_2 = 10$ towards a stable equilibrium. The figure for *Zeiraphera* is shown once the population has settled down from its initial conditions.

dynamics (Chapter 5). Hassell et al. classified the larch bud-moth in the Engadine Valley, Switzerland, as stable, although this population, was, according to Turchin and Taylor 'arguably the most convincing example of a . . . cyclical system in our data set'. For further insights into the mechanisms underpinning cyclical and other population dynamics see Turchin (2003).

6.3 Competition models

In Chapter 5 we considered the possibility of *intra*specific competition as a mechanism producing density dependence. In this chapter our attention is on competition between species, or *inter*specific competition. This will serve as a prelude to a generalized description of interactions between species in Chapter 7.

Much of the classic work on interspecific competition involved insects and micro-organisms including beetles in stored grain (Crombie 1945, 1946, 1947, Park et al. 1964), aquatic protozoa (Gause 1934, 1935) and yeast (Gause 1932, 1934). For example, the experiments of Crombie (1945, 1946) showed how three combinations of the beetles *Tribolium confusum* and *Oryzaephilus surinamensis* converged to the same population equilibrium. The beetles were fed on wheat, presented as either cracked grain or flour. In cracked wheat each species cultured in isolation increased to between 420–450 adults in 150 days (represented as carrying capacities K_1 and K_2 in Table 6.2). The change in numbers over time is plotted on the phase plane in Fig. 6.9. However, when the species were combined, *Tribolium* reached 360 individuals and *Oryzaephilus* 150 individuals; thus the total was greater than the species in monoculture (represented as N_1^* and N_2^* in Table 6.2). The results were independent of the initial number of beetles, indicating that this was a globally stable equilibrium. As Pontin (1982) noted, 'the total number of beetles in mixed culture at equilibrium is equal to or greater than the carrying capacity number of either (species) alone so the combination may be more efficient at converting grain to beetles.' Changing the medium from flour grains to fine flour resulted in the extinction of *Oryzaephilus*. Small pieces of tubing which provided shelter for *Oryzaephilus* allowed the latter to survive under fine-flour conditions (Table 6.2). The same general effects of habitat structure on coexistence were found in the predator–prey experiments of Huffaker (1958).

Tribolium individuals are carnivorous: the larvae and adults eat eggs and pupae of their own species and also those of *Oryzaephilus*. Adult *Oryzaephilus* also consume *Tribolium* eggs but at a lower rate than its own are consumed by *Tribolium*. Therefore this interaction is part predation and part competition for resource. This emphasizes the need for a generalized model in which a variety of interactions are encompassed.

Table 6.2 Values of carrying capacity of *Tribolium confusum* (species 1) and *Oryzaephilus surinamensis* (species 2) in isolation (K_1 and K_2) and at equilibrium in mixtures (N_1^* and N_2^*) estimated from experiments. Values for the competition coefficients (β_{12}, β_{21}) are derived in the text (data in Pontin 1982 from Crombie 1946).

| | Equilibrium numbers | | | | Competition coefficients | |
| | Alone | | Together | | | |
	K_1	K_2	N_1^*	N_2^*	β_{12}	β_{21}
Cracked wheat	425	445	360	150	0.4	0.8
Fine flour, 1 mm tubes	175	400	175	80	Small	1.8

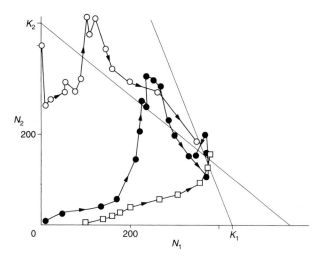

Fig. 6.9 Population trajectories from three competition experiments started with different numbers of *Tribolium* (N_1) and *Oryzaephilus* (N_2) in cracked-wheat cultures. The lines represent zero-growth isoclines of each species (from Crombie 1946, reprinted in Pontin 1982).

A pair of competing species such as *Tribolium* and *Oryzaephilus* can be described by a pair of simultaneous nonlinear differential equations similar to those used for the predator–prey interactions by Lotka and Volterra and, indeed, developed by Volterra (1926, 1928) and Gause (1934):

$$dN_1/dt = r_1N_1(K_1 - N_1 - \beta_{12}N_2)/K_1 \tag{6.10}$$

$$dN_2/dt = r_2N_2(K_2 - N_2 - \beta_{21}N_1)/K_2 \tag{6.11}$$

where N_1 and N_2 are the densities of the competing species. β_{12} describes the fraction of species 1 converted into species 2. This is known as the competition coefficient and is generalized to interactions between species i and j as β_{ij}. r is the intrinsic rate of change for a given species and K_1 and K_2 are the carrying capacities of species 1 and 2 in isolation. Note that equation 6.10 could also be written as:

$$dN_1/dt = r_1N_1((K_1 - (N_1 + \beta_{12}N_2))/K_1$$

As β_{12} is the fraction of species 1 converted to species 2 then $N_1 + \beta_{12}N_2$ is effectively the density of N_1, replacing N_1 in the ordinary logistic equation (Chapter 5).

Investigation of stability with the phase plane begins, as with the predator–prey system, with the zero-growth isoclines ($dN_1/dt = 0$ and $dN_2/dt = 0$). We will go through this more rapidly than the predator–prey example as many of the principles are the same. If we take the example of *Tribolium* and *Oryzaephilus* we find the values of K_1, K_2 and N_1^* and N_2^* (the equilibrium densities in mixtures) from the experiment (Table 6.2). With zero growth, equations 6.10 and 6.11 are:

$$0 = r_1 N_1^*(K_1 - N_1^* - \beta_{12} N_2^*)/K_1 \tag{6.12}$$

$$0 = r_2 N_2^*(K_2 - N_2^* - \beta_{21} N_1^*)/K_2 \tag{6.13}$$

From equation 6.12 we have either $r_1 N_1^* = 0$ (the trivial solution) or $K_1 - N_1^* - \beta_{12} N_2^* = 0$, which gives $N_1^* = K_1 - \beta_{12} N_2^*$. Similarly, equation 6.13 yields $N_2^* = K_2 - \beta_{21} N_1^*$. As the values of N_1^*, N_2^*, K_1 and K_2 are known we can use these equations to find β_{12} and β_{21} and plot the zero-growth lines (Figs 6.9 and 6.10).

The outcomes of competition can now be investigated on the phase plane in the same way as for predator–prey interactions. These outcomes are (i) stable competition in which both species coexist, (ii) unstable competition in

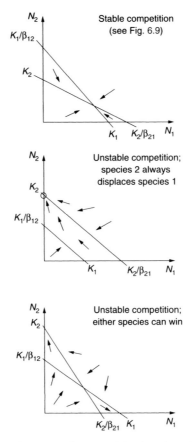

Fig. 6.10 Three outcomes of interspecific competition based on three combinations of zero growth isoclines from equations 6.12 and 6.13.

which one species always displaces the second species – that is, there is a fixed hierarchy of competition – and (iii) unstable competition in which either species can win (Fig. 6.10). The assessment of competition coefficients in the field is developed in the context of generalized Lotka–Volterra models in Chapter 7. Another interesting development is the use of Lotka–Volterra competition models in combination with Markov models (Spencer & Tanner 2008)

Other competition models have been developed using discrete-time versions of equations 6.10 and 6.11; for example, the model of Hassell and Comins (1976):

$$N_{1,t+1} = \lambda_1 N_{1,t}(1 + a_1(N_{1,t} + \beta_1 N_{2,t}))^{-b_1}$$

$$N_{2,t+1} = \lambda_2 N_{2,t}(1 + a_2(N_{2,t} + \beta_2 N_{1,t}))^{-b_2}$$

β_1 and β_2 are the competition coefficients, a_1 and a_2 give the threshold densities at which density dependence begins and b_1 and b_2 are parameters that describe the different types of intraspecific competition, with extremes of scramble and contest (see Chapter 5 for discussion of the one-species analogue of this model). λ is the finite rate of population change for each species. Atkinson and Shorrocks (1981) used the model of Hassell and Comins to explore the effect of aggregation of competing *Drosophila* species on coexistence. The degree of aggregation was modelled using the negative binomial distribution. An increased aggregation of the superior competitor promoted coexistence. The importance of aggregation is highlighted in the next section with respect to host–parasitoid models.

6.4 Models of host–parasitoid interactions

Although delayed density dependence may arise due to interactions between herbivores and plant (section 6.2) this does not mean that interactions with the food plant are always the reason for the cycles in herbivores. Delayed density dependence may also occur due to interactions between a herbivore and its natural enemies; for example, parasitoids. Parasitoids are primarily wasps which lay their eggs on or inside a host larva, such as a moth or a fly. Indeed, Varley originally coined the term delayed density dependence because of his work on the parasitoids of the herbivorous fly, *Urophora jaceana*, which feeds on black knapweed, *Centaurea nigra*.

It has been estimated that more than 10% of all metazoan animals are parasitoids (Hassell & Godfray 1992) so it is important to understand how parasitoids interact with their hosts and, in particular, the type of dynamics which may be produced. Certain characteristics of parasitoid behaviour and life history may affect their dynamic interaction with their hosts (usually herbivorous insects, although there are parasitoids of parasitoids called hyperparasitoids) including (i) searching area of the female parasitoid and (ii) interactions between female parasitoids.

The earliest model of host–parasitoid interactions was constructed by Nicholson and Bailey (1935). Their model was important as it made the case for time delays rather than the continuous time Lotka–Volterra equations described above. The main assumptions of their model are given below and illustrated in Fig. 6.11.

1 Either zero or one parasitoid is produced per host (even if more than one egg is laid on a host).
2 Each female parasitoid searches an area a, finding all the hosts. Therefore the probability of a host being attacked is a/A where A is the total area and the probability of not being parasitized is $1 - a/A$. If P is the density of parasitoids and A is the total area then there are AP female parasitoids. If parasitoids search independently and at random then the probability of a host not being attacked by any parasitoid is $(1 - a/A)^{AP}$. If a/A is replaced by α (defined as the proportion of total hosts encountered by one parasitoid per unit time) then we have $(1 - \alpha)^{AP}$ which is equal to $e^{-\alpha P}$, the first term of the

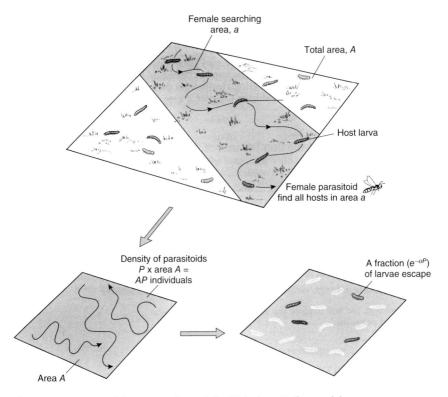

Fig. 6.11 Summary of the assumptions of the Nicholson–Bailey model.

Poisson distribution; that is, the probability of the host not containing parasitoids. The probability of being attacked at least once is then 1 – (probability
of not being attacked), or $1 - e^{-\alpha P}$.

3 The finite rate of increase of hosts (λ_H, in the absence of parasitoids) and
parasitoids (λ_P) and the density of hosts (H) are known.

Using the above reasoning and assumptions Nicholson and Bailey produced
a set of two equations in discrete time, one for the dynamics of the host (H)
and one for the dynamics of the parasitoid (P):

$$P_{t+1} = \lambda_P H_t \left(1 - e^{-\alpha P_t}\right) \tag{6.14}$$

$$H_{t+1} = \lambda_H h_t e^{-\alpha P_t} \tag{6.15}$$

Thus the number of parasitoids at time $t + 1$ (P_{t+1}) is equal to the number of
hosts at time t (H_t) multiplied by the fraction of hosts which are attacked
$(1 - e^{-\alpha P_t})$ multiplied by the finite rate of increase of the parasitoids (λ_P);
whereas the number of hosts at $t + 1$ (H_{t+1}) is equal to the number of hosts
at t (H_t) multiplied by the fraction of hosts which are not attacked ($e^{-\alpha P}$)
multiplied by the finite rate of increase of the host (λ_H).

The dynamics produced by these two equations have an unstable equilibrium and therefore produce cycles which can very easily become divergent
(Fig. 6.12a) and lead to the local extinction of the parasitoid.

One option to stabilize the host populations is to introduce host (prey)
density dependence. Host stability can be achieved by multiplying the finite
rate of increase of the host (λ_H) by a linear term $1 - H_t/K$, which can represent
intraspecific competition among hosts. Beddington et al. (1975) examined
the effect of including host density-dependent regulation. Stability was indeed
much more likely and new types of dynamics were produced, such as five
and 20 point cycles. This was compared with the host equation in the absence
of predation which followed the standard period-doubling route to chaos
(Chapter 5).

Other possibilities exist for stabilizing the host–parasitoid system; for
example, the two assumptions of fixed search area and random searching by
parasitoids have been shown to be unrealistic and to affect the stability of
the interaction (Hassell & Godfray 1992). The effect of search area on stability
was demonstrated by assuming competition between the parasitoids (Hassell
& Varley 1969, Hassell & May 1973), which produces cycles or a stable equilibrium dependent on the intensity of interference (Fig. 6.12b). The effect of
an aggregated distribution of parasitoids on host–parasitoid stability has been
modelled using the negative binomial (May 1978). This will also produce a
stable equilibrium or cycles dependent on the degree of aggregation.

The latter is an example of positive spatial density dependence in which,
at one point in time, densities of hosts are spatially variable and high host
densities receive an increased rate of parasitism. The density dependence
discussed in previous chapters was temporal density dependence where

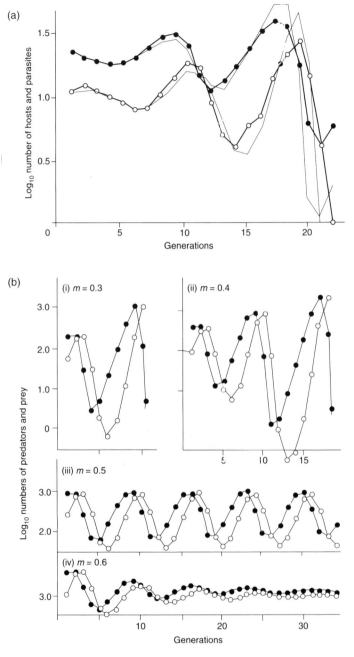

Fig. 6.12 Dynamics of host and parasitoid (a) without density-dependent regulation of host or parasitoid. Observed fluctuations from an interaction between the greenhouse whitefly, *Trialeurodes vaporariorum* (closed circles) and a parasitoid wasp *Encarsia formosa* (open circles). Thin lines show results of the Nicholson–Bailey model. (b) Model with increasing levels of competition between parasitoids. Parasitoid (open circles) and prey (closed circles) oscillations from a modified Nicholson–Bailey model showing the progressive stability as the interference constant (m) increases from 0.3 to 0.6 (Hassell & Varley 1969). Figure reprinted from Hassell (1976).

densities fluctuated between years or other time periods. Field populations may be expected to show both temporal and spatial density dependence. However, it is not always the case that parasitoids show aggregation in response to local host density. The degree of parasitoid aggregation was determined from field data relating to the successful biological control of California red scale insects by their parasitoids (Reeve & Murdoch 1985). In this case no evidence was found for parasitoid aggregation at any spatial scale, despite the fact that it had been believed to stabilize the interaction. This was important because local extinction of the host is undesirable as the parasitoid becomes extinct too and therefore the biological control agent will not establish. The ideal end point to biological control is that both the host and the control agent remain at low equilibrium densities, as illustrated by the control of the prickly pear, *Opuntia* sp., by the moth *Cactoblastis cactorum* in Australia (Krebs 1994). Although lauded as a major success, Raghu and Walton (2007) consider that it may be atypical of biological control and cite the enormous effort involved in distributing the 2 billion egg sticks (each stick containing approximately 50–100 eggs) of *C. cactorum* across the infested region of eastern Australia.

In Chapter 8 the role of spatial scale is explored with the basic Nicholson–Bailey model and the consequences for dynamics and stability examined. As a conclusion to this section and looking forward to Chapter 8, it is interesting to note that Nicholson and Bailey recognized many of the possible developments of their model, leading to persistence. This included regulation of the host (as above):

> When the density of a species becomes very great as a result of increasing oscillation the retarding influence of such factors as scarcity of food or of suitable places to live is bound to be felt. Clearly these factors will prevent unlimited increase in density so . . . that the oscillation is perpetually maintained at a large constant amplitude in a constant environment.

and the anticipation of a spatial element:

> A probable ultimate effect of increasing oscillation is the breaking up of the species-population into numerous small widely separated groups which wax and wane and then disappear, to be replaced by new groups in previously unoccupied situations.
> Nicholson and Bailey (1935)

6.5 An application of predator–prey models: harvesting populations

Much of our understanding of the harvesting of wild populations has come from the fisheries literature, with classic long-term studies in the North Sea, Atlantic and Pacific. These studies have used both continuous- and discrete-time population models. These have been complemented by studies on terrestrial populations including high-profile animals such as the African elephant. In this section we will consider how population models can be used to determine appropriate levels of harvesting. We will start with an unstructured model in continuous time and then explore how much more can be learned from a structured model.

6.5.1 Unstructured population models

Assume that population growth is continuous and described by the logistic equation. Although in many cases the logistic equation is too simple a description of population change (see discussion in Chapter 5), it provides a useful entry point to understanding the dynamic possibilities of harvesting. We begin by plotting the rate of population growth, dN/dt, against population size, N_t (Fig. 6.13).

You should note that the curve in Fig. 6.13 is a parabola, which is the shape generated by a quadratic equation (recall that the right-hand side of the logistic equation in expanded form is $rN_t - rN_t^2/K$). All population sizes which yield values of $dN/dt > 0$ can, in theory, be subject to harvesting. This means that a fraction of any growing population should be able to be harvested without causing the population to become extinct. The maximum sustainable yield occurs where the maximum value of dN/dt occurs; that is, when the population is growing most rapidly. This result is generally true for all population growth curves (see the discussion in May & Watts 1992).

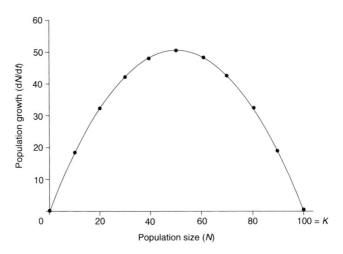

Fig. 6.13 Population growth dN/dt described by the logistic equation plotted against population size (N). K is the carrying capacity.

To determine the population density corresponding to the maximum sustainable yield we note that the maximum occurs when the gradient of the curve equals 0:

$$d(dN/dt)dN = 0 \qquad\qquad (6.16)$$

We will drop the subscript t to make the equations easier to read (N will still mean the population density at time t). For the logistic curve in Fig. 6.13 the derivative of the expression $rN - rN^2/K$ with respect to N is $r - 2rN/K$. Therefore

$$r(1 - 2(N/K)) = 0$$

The solution to this equation is either $r = 0$ (a trivial solution) or $(1 - 2(N/K)) = 0$. Rearranging the latter yields the solution that $N = K/2$. So the population is growing at its fastest when it is half its maximum density. This agrees with the shape of the logistic curve where the point of inflexion (the maximum gradient) was at $K/2$ (Figs 5.11 and 6.13).

To find the maximum value of dN/dt we need to substitute the size at which the maximum occurs ($K/2$) for N in the logistic equation $dN/dt = rN(1 - N/K)$. This gives $dN/dt = rK/4$. This method of finding the maximum value is useful if the function for population growth is more complex than the logistic equation.

To develop these ideas let us assume a general form of population growth equation $dN/dt = f(N)$ which describes the rate of change in population size as some function (f) of population size. If this function $f(N)$ has a maximum value or values of dN/dt over a certain range of values of N then differentiation can be used to find the maximum value. You should note that the criterion of $d(dN/dt)dN = 0$ (equation 6.16) is not sufficient to identify a maximum value. It might equally identify a minimum value or a point of inflexion. A harvesting function can be incorporated into the general population growth equation (Beddington 1979):

$$dN/dt = f(N) - h(N) \qquad\qquad (6.17)$$

where $h(N)$ gives the reduction in dN/dt at a particular value of N due to harvesting. The population change and harvesting functions of the generalized form of equation 6.17 can be combined on one graph as they are both functions of N (Figs 6.14 and 6.15). We will now consider several harvesting possibilities with a logistic growth curve.

If $h(N)$ is constant and therefore independent of N, harvesting is represented by a horizontal line on the graph. We know that sustainable harvesting can only occur when $dN/dt > 0$; when $f(N) > h(N)$ in equation 6.17. The area on the graph when this condition is met is shaded in Fig. 6.14 and lies between $N = 20$ and $N = 80$. What happens to the population at different population sizes? For example, consider a population of size 70. Here $f(N) > h(N)$ and therefore population size increases ($dN/dt > 0$, equation 6.17).

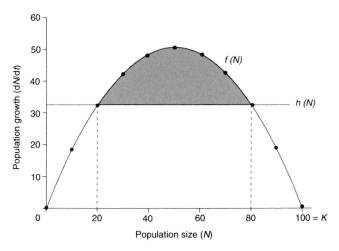

Fig. 6.14 Constant harvesting $h(N)$ and logistic population growth $f(N)$ plotted against N (equation 6.17). The shaded area shows where $f(N) > h(N)$. Where the two lines (functions) intersect the rate of population change is 0.

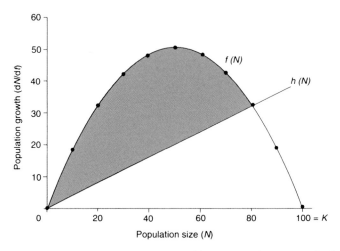

Fig. 6.15 Linear increase in harvesting $(h(N))$ and logistic population growth $(f(N))$ plotted against population size (N).

In other words, from population size 70 we move to the right on the graph. Conversely, for a population of size 90, $f(N) < h(N)$ and so $dN/dt < 0$ and the population size decreases.

By this reasoning it can be seen where $h(N) = f(N)$ at $N = 80$ (and therefore $dN/dt = 0$) there is a locally stable equilibrium point. A population which receives a small displacement away from that equilibrium point will tend to return to it. What then of the other point at which $dN/dt = 0$, at $N = 20$? If

$N < 20$ then $f(N)$ is less than $h(N)$ and therefore $dN/dt < 0$. So, the population will continue to decrease if reduced below 20 until it reaches $N = 0$, which is local extinction ($N = 0$ is effectively a third equilibrium point, which is locally stable). If the population size is increased above $N = 20$ then $f(N) > h(N)$ and therefore the population size will continue to increase until it reaches the stable equilibrium point at $N = 80$.

Even the very simple scenario of constant harvesting combined with logistic growth provides the dynamic possibilities of extinction (below $N = 20$) and local stability (at $N = 80$). A second possibility for the harvesting function is that it increases linearly with prey population size (Fig. 6.15), so that the more fish there are, the more people go fishing. Again, $N = 80$ is seen as a stable equilibrium. In both this and the previous example, if the prey population is pushed beyond K (100) then it is predicted from the model that it will return towards K, but it will not stay there as $h(N)$ is still greater than $f(N)$ and therefore it continues to return to 80.

6.5.2 Structured population models

We begin by considering an unstructured population in discrete time in which the population dynamics are described by λ, the finite rate of population change, which is equivalent to the dominant eigenvalue of the population projection matrix for a structured population. We know that a population with $\lambda < 1$ will decline in numbers, so if a harvesting policy is to be sustainable it should not decrease the value of λ below 1. It is therefore possible to arrive at a simple definition of the maximum amount of a population which can be harvested. For example, if $\lambda = 2$, the maximum amount which can be removed is that which keeps $\lambda = 1$:

Fraction harvested $= (\lambda - 1)/\lambda = 1/2$

If $\lambda = 3$, the maximum fraction of the population which could be harvested is $2/3$. This assumes that λ is constant from year to year and that a constant fraction is removed in each time period.

Now consider the implications of age, size or stage structure. To explore the effect of population structure on harvesting consider the two-stage model of biennial plants used previously (equations 4.9 and 4.10):

$$\begin{pmatrix} R \\ F \end{pmatrix} = \begin{pmatrix} 0 & fs_{0,1} \\ s_{1,2} & 0 \end{pmatrix} \begin{pmatrix} R \\ F \end{pmatrix}$$

$$\mathbf{v}_{t+1} \qquad\qquad \mathbf{M} \qquad\quad \mathbf{v}_t$$

The characteristic equation was $\lambda^2 = fs_{0,1}s_{1,2}$ (equation 4.17). We know that the maximum fraction which can be harvested is given by $(\lambda - 1)/\lambda$, so it becomes of interest to see how manipulation of f, $s_{0,1}$ and $s_{1,2}$ affects $(\lambda - 1)/\lambda$.

Assume that a fraction m_1 of flowering plants is harvested prior to setting seed and therefore the fraction of surviving plants is represented by the fraction $(1 - m_1)$. Note that removal of flowering plants after seed set for a monocarpic species is not going to affect the population dynamics. This harvesting mortality can be incorporated into the model as follows:

$$\begin{pmatrix} R \\ F \end{pmatrix} = \begin{pmatrix} 0 & fs_{0,1} \\ s_{1,2}(1 - m_1) & 0 \end{pmatrix} \begin{pmatrix} R \\ F \end{pmatrix}$$

$\mathbf{v}_{t+1}$ **M** with harvesting $\mathbf{v}_t$
             of F

In a similar way, we might imagine that a fraction m_2 of rosette plants is harvested (of course, either m_1 or m_2 or both can have zero values). This can also be included in the model:

$$\begin{pmatrix} R \\ F \end{pmatrix} = \begin{pmatrix} 0 & fs_{0,1}(1 - m_2) \\ s_{1,2}(1 - m_1) & 0 \end{pmatrix} \begin{pmatrix} R \\ F \end{pmatrix}$$

$\mathbf{v}_{t+1}$ **M** with harvesting $\mathbf{v}_t$
             of F and R

(6.18)

We can now determine λ for the new model with the harvesting mortalities. The characteristic equation of the matrix equation 6.18 is:

$$\lambda^2 = fs_{0,1}(1 - m_2)s_{1,2}(1 - m_1)$$

(6.19)

This is obviously similar to the characteristic equation without harvesting. Indeed, if we define λ_u as λ when the population is unharvested and λ_h as λ when the population is harvested, then we can derive a useful result from equation 6.19. Taking the square root of that equation:

$$\lambda_h = \pm\sqrt{(fs_{0,1}s_{1,2})r[(1 - m_2)(1 - m_1)]}$$
$$\text{As } \lambda_u = \pm\sqrt{(fs_{0,1}s_{1,2})}$$
$$\lambda_h = \pm\lambda_u\sqrt{[(1 - m_2)(1 - m_1)]}$$

Thus the value for λ_h is given by the original $\lambda(\lambda_u)$ multiplied by the square root of $(1 - m_2)(1 - m_1)$. Using matrices we can approach the problem from a slightly different angle to shed more light on the expression $\sqrt{(1 - m_2)(1 - m_1)}$.

Let us factorize the square matrix from equation 6.18:

$$\begin{pmatrix} 0 & fs_{0,1}(1 - m_2) \\ s_{1,2}(1 - m_1) & 0 \end{pmatrix}$$

$$\begin{pmatrix} 0 & fs_{0,1} \\ s_{1,2} & 0 \end{pmatrix} \begin{pmatrix} 1 - m_1 & 0 \\ 0 & 1 - m_2 \end{pmatrix}$$

The left-hand matrix is the original transition matrix for the unharvested population and has an eigenvalue of λ_u. The right-hand matrix is composed of the two harvesting 'survivals' and can therefore be referred to as a harvesting matrix (Lefkovitch 1967). We could use this method for any structured population to explore the effects of harvesting of different ages, sizes or stages on population dynamics. Whereas matrix models can be valuable for analysis of the effects of harvesting and other management we need to be cautious as variations across time and space may alter λ and possibly reduce the maximum sustainable yield well below its theoretical value. For this reason, modellers have increasingly considered the role of stochasticity in harvesting models (Caswell 2000b). Sensitivity and elasticity analyses (Chapter 4) also help in interpretation of effects of harvesting.

Matrix models are now used routinely in investigating the effects of harvesting on structured populations. For example, Olmstead and Alvarez Buylla (1995) have explored the possibility of sustainable harvesting of two tropical palm species using matrix models. They calculated the population growth rates from stage-structured models and estimated the amount of adult trees which could be harvested per unit area. They concluded that only one species could be harvested as the λ of the other species was only 1.05. Similarly, a study of harvesting of *Quercus* in Mexico confirmed that removal of just 5% of adults could cause population decline (Alfonso-Corrado et al. 2007). Other studies have been applied to organisms as diverse as capybara (the world's largest rodent; Federico & Canziani 2005) and red coral (Santangelo et al. 2007). Inevitably, such studies also have conservation applications as harvesting threatens the viability of a wide range of species.

CHAPTER 7

Community models

7.1 Introduction to modelling of ecological communities

So far the emphasis has been on population dynamics of a single species or clade and two interacting species such as predators and prey or competitors. These pairwise interactions rarely occur in isolation. In this chapter we consider sets of interactions between an assemblage of species in a given locality, the dynamics of an ecological community. The evolutionary equivalent has been addressed briefly in terms of the diversity-dependent reductions in diversification rate. We will be asking similar questions of an ecological community as we did of a population; in particular we will be interested in the long-term stability of the community. In addition to the density of the population(s) a new variable arises in community dynamics: the diversity of the community. Although diversity is a major topic in its own right, for our purposes we simply need to know that the diversity of ecological has two components: species richness (the number of species) and the relative abundance of species; that is, their evenness (or not). It is these two components that are incorporated into diversity indices such as the Simpson and Shannon–Weiner indices.

The modelling of community dynamics has three related problems, which will be tackled in this chapter. First, there is a need to consider the full range of interactions between species (Chapter 6), described by all combinations of 0/+/– where 0 represents no interaction, + is a positive (beneficial) effect of species A on B and – is a detrimental effect of A on B (Fig. 7.1). The second problem is that we need to assign a strength – that is, a magnitude – to these interactions. The sign and magnitude of the interactions will be combined in a community matrix. Matrix methods will be used to investigate the stability and dynamics of communities. The final problem is the large number of species that may be present in a community. In response to this we will see how species can be objectively removed from the model to move towards the aim of the simplest realistic model.

These models of ecological communities will mostly assume a pool of species within a given area, each of which can change in abundance over time (including becoming locally extinct) but do not assume any immigration or possible replacement of species as seen during successional change.

		Effect of species j on i		
		+	0	−
Effect of species i on j	+	++	+0	+−
	0	0+	00	0−
	−	−+	−0	−−

Fig. 7.1 Classification of interactions. Signs and magnitudes of a_{ij} are considered in section 7.2.

In addressing the structure and dynamics of large communities we should recognize that even the addition of one species to the two-species interactions of Chapter 6 produces important changes in the dynamics of the component species (Gilpin 1979, Guckenheimer & Holmes 1983, Schaffer 1985). An investigation of a food-chain model composed of just three species revealed that chaotic dynamics occurred within biologically feasible parameter values, whereas chaos was not possible in two species models (Hastings & Powell 1991).

7.2 Food-web structure and the community matrix

7.2.1 Introduction

Measures of complexity in ecological communities are often based on the composition of the community represented by the number and relative abundance of species. These measures, which include the myriad diversity indices, do not tell us very much about the functioning of communities and their properties, such as stability. To quantify functional aspects of the community and understand how they may respond to perturbations, we need to quantify the interactions between species. The potential complexity of such interactions in real ecological communities is illustrated by the food webs in Fig. 7.2.

The food web can be characterized by the number of component species and the average number of links between them. The latter is known as linkage density. This idea can be extrapolated to other interactions between species, such as mutualisms. A motivation of this work has been to understand how communities with different linkage densities might respond to environmental perturbations, for example pollution events. A naïve view of this suggests two alternative outcomes. A highly connected community (high linkage density) might be predicted to be more resistant to perturbations, for example because any one species will have many possible food sources. Alternatively we might imagine that a highly connected community will be less resistant because harmful effects on just a few species will potentially

(a)

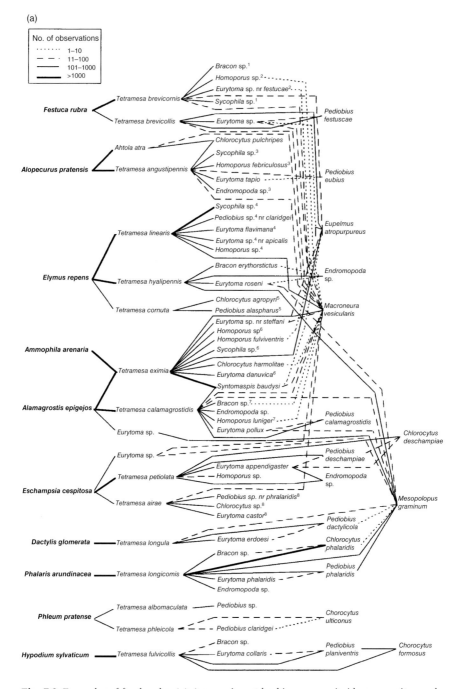

Fig. 7.2 Examples of food webs. (a) A grass–insect herbivore–parasitoid community used to demonstrate the effects of sampling effort (Martinez et al. 1999). (b) A diagrammatic view of a salt-marsh food web illustrating the complexity with inclusion of parasites (Lafferty et al. 2006, 2008). Parasites are light-coloured and free-living species are dark-coloured.

(b)

Fig. 7.2 *Continued*

Table 7.1 A matrix of interaction coefficients for a hypothetical four-species community.

	On species . . .	Interaction coefficient			
		A	B	C	D
Effect of species . . .	A	−1.0	−0.5	0	+1.0
	B	+0.5	−1.0	+0.5	0
	C	0	+0.5	−1.0	+0.8
	D	+1.0	0	−0.8	−1.0

cause major problems throughout the community. It is these types of questions that we will address in this chapter.

One way of describing the strength of interactions in a community is to construct a matrix of interaction coefficients. For example, a hypothetical four-species community can be represented as shown in Table 7.1. The representation can include both feeding and non-feeding interactions. For example, the effect of species A on species B is negative with a strength of 0.5. The intraspecific interactions (e.g. A on A) are all negative. Some of the interactions are symmetric, for example a positive effect of A on D and D on A, which is a fully mutualistic interaction. A value of 0 indicates no link between the species.

The analysis of ecological community stability is part of a wider debate about the stability or not of complex systems. Gardner and Ashby (1970) asked whether large systems (biological or otherwise) which were assembled

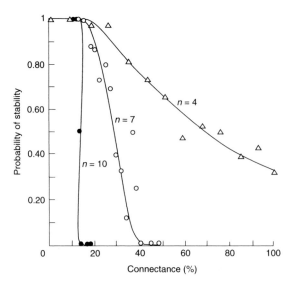

Fig. 7.3 The relationship between stability, connectance and component richness. A breakpoint or threshold of stability is observed at higher numbers of components. From Gardner and Ashby (1970).

at random would be stable. In their words, they were concerned with 'an airport with 100 planes, slum areas with 10^4 persons or the human brain with 10^{10} neurons . . . [where] stability is of central importance'. Forty years on and these questions are no less pertinent; indeed, we may add an order of magnitude to some of these problems and present entirely new problems for solution; for example, global access to the Internet. Gardner and Ashby showed that for small numbers of components (n; e.g. neurons or species) stability declines with connectance between components (Fig. 7.3). As the number of components increases, the system moves rapidly to a breakpoint situation when a small change in connectance will result in a switch from stability to complete instability.

These results were developed by May (1972) in an ecological context. He concluded that increased numbers of species do not automatically imply community stability and in fact may produce just the opposite effect (see Jansen & Kokkoris 2003 and references therein for further debate on these results). Increased stability with complexity was promoted by Elton (1958), whose conclusions were partly based on detailed case studies of invasive species, such as the giant snail *Achatina fulica* into Hawaii and the red deer *Cervus elaphus* into New Zealand, which contributed to dramatic declines in the endemic species of those islands (see May 1984 and Pimm 1984 for a critique of Elton's views). May also demonstrated that a species that interacts widely with many other species (high connectance) does so weakly (small interaction coefficient) and, conversely, those that interact strongly with others do so with a smaller number of species. He predicted that communities which are compartmentalized into blocks (effectively communities within communities) may be stable while the whole may not be. These ideas were explored further by Tregonning and Roberts (1979) who examined the stability of a randomly constructed model community in which the interaction coefficients were non-zero and the values chosen randomly. They began by running the model with 50 species, and used two methods of species elimination: a species was either chosen at random or the species with the most negative equilibrium value selected. Therefore, in the latter case they removed the most ecologically unrealistic species, as all species needed to have a positive equilibrium value. This process was continued until all species had a positive equilibrium value. This was defined by Tregonning and Roberts as the homeostatic system: one that was ecologically feasible and at equilibrium. Under selective removal the mean number of species comprising a homeostatic system was 25 and the largest 29. However, if elimination was random then the largest homeostatic system was 4 and the mean 3.3. Further understanding of these results requires a deeper insight into the nature of the community matrix used by May, Tregonning and Roberts and others. This is the subject of the next section.

7.2.2 Construction and stability of the community matrix

Levins (1968) first devised a matrix of Lotka–Volterra competition coefficients to describe community structure and predict community stability. This was a multi-species version of the two-species competition model (described earlier). This idea was developed by May (1972, 1973a) giving a general version of the matrix called the community matrix (sometimes referred to as the stability matrix) which expressed the effect of species j on species i near equilibrium. The community matrix makes some assumptions about the dynamics of its constituent species; in particular, it assumes that prey and competitors will be regulated so that in the absence of any interspecies interactions they will return to equilibrium. It is also assumed that predators decline exponentially in the absence of prey.

   The community matrix allows insights into the stability of the community, with the dynamics of each species described by a nonlinear first-order differential equation. The aim is to create a matrix $\mathbf{M}$ of all interactions between the species in the community. It is assumed that species will be regulated so that, in the absence of any interspecies interactions, they will return to equilibrium. The following notation will be used:

α_{ij} the interaction coefficients between species i and j (expressed as the effect of species j on the growth rate of species i), including α_{ii}, the intraspecific interaction; the magnitude of α ranges from 1 to 0, which represents no interaction (the sign of α is considered below);

N_i density or biomass of species i;
r_i the intrinsic rate of change of species i;
s the number of species.

A generalized Lotka–Volterra model following Roberts (1974) and Tregonning and Roberts (1979) and referred to by them as the multi-species quadratic model, is summarized for any number of species by:

$$dN_i/dt = N_i\left(r_i + \sum_{i=1}^{s}\alpha_{ij}N_j\right) \tag{7.1}$$

where r_i is positive for a producer (prey, competitor) and negative for a consumer (predator, pathogen) following the convention in Chapter 6. Therefore consumers decline exponentially in the absence of producers. Producers show density-dependent regulation as illustrated by the one-species version of equation 7.1:

$$dN_1/dt = N_1(r_1 + \alpha_{11}N_1)$$

or

$$dN_1/dt = r_1N_1 + \alpha_{11}N_1^2 \tag{7.2}$$

This equation is equivalent to the logistic equation $dN/dt = r_1N_1 - r_1N_1^2/K$ with α_{11} equal to $-r_1/K$. Equation 7.2 shows why Tregonning and Roberts referred to the model as the (multi-species) quadratic model. In Chapter 5 it was shown that populations described by the logistic equation had a stable equilibrium of K. The equilibrium occurs at $dN/dt = 0$; therefore, for equation 7.2:

$$0 = rN_1^* + \alpha_{11}N_1^{*2}$$

where N_1^* is the equilibrium population size. Factorize the right-hand side to give $N_1^*(r + \alpha_{11}N_1^*)$ and rearrange the non-trivial solution $(r + \alpha_{11}N_1^*) = 0$ to give:

$$N_1^* = -r/\alpha_{11}$$

As r must be positive for a single producer species, α needs to be negative to give a positive value of N_1^*. The sign of α is important and we will return to it later. (Note that the trivial solution is $N^* = 0$.)

With two species, equation 7.1 gives:

$$dN_1/dt = N_1r_1 + \alpha_{11}N_1N_1 + \alpha_{12}N_2N_1 \tag{7.3}$$

$$dN_2/dt = N_2r_2 + \alpha_{21}N_1N_2 + \alpha_{22}N_2N_2 \tag{7.4}$$

We can compare the parameters α_{11}, α_{12}, α_{21} and α_{22} with the parameters in the competition and predator–prey equations in Chapter 6. In comparison with the competition equations, $\alpha_{11} = -r_1/K_1$, $\alpha_{12} = -\beta_{12}r_1/K_1$, $\alpha_{21} = -\beta_{21}r_2/K_2$

and $\alpha_{22} = -r_2/K_2$. Generalizing for interactions between species i and j, $\alpha_{ij} = -\beta_{ij}$ r_i/K_i and substituting α_{ii} for $-r_i/K_i$, $\alpha_{ij} = \beta_{ij}\alpha_{ii}$; that is, the competition coefficient multiplied by the intraspecific interaction coefficient. Compared with the predator–prey equations 6.3 and 6.4 (assuming N_1 is prey and N_2 is predator): $\alpha_{11} = -r_1/K_1$, $\alpha_{12} = -a$, $\alpha_{22} = 0$ and $\alpha_{21} = b$. Also r_2 will be negative and r_1 will be positive.

We can therefore see how, with different values of r and α_{ij}, equation 7.1 can provide a generalized description of Lotka–Volterra dynamics covering interactions such as competition and predation.

The community matrix is then derived by considering the community at equilibrium. If we take the two-species example:

$$0 = N_1{}^*r_1 + \alpha_{11}N_1{}^*N_1{}^* + \alpha_{12}N_2{}^*N_1{}^* \tag{7.5}$$

$$0 = N_2{}^*r_2 + \alpha_{21}N_1{}^*N_2{}^* + \alpha_{22}N_2{}^*N_2{}^* \tag{7.6}$$

The species densities can be evaluated at equilibrium:

$$-r_1 = \alpha_{11}N_1{}^* + \alpha_{12}N_2{}^*$$

$$-r_2 = \alpha_{21}N_1{}^* + \alpha_{22}N_2{}^*$$

In matrix form this is

$$\begin{pmatrix} -r_1 \\ -r_2 \end{pmatrix} = \begin{pmatrix} \alpha_{11} & \alpha_{12} \\ \alpha_{21} & \alpha_{22} \end{pmatrix} \begin{pmatrix} N_1{}^* \\ N_2{}^* \end{pmatrix} \tag{7.7}$$

The values of $N_1{}^*$ and $N_2{}^*$ can be calculated using matrix algebra by finding the inverse of the matrix of coefficients and multiplying both sides to give:

$$N_1{}^* = \left(\frac{-\alpha_{22}r_1 + \alpha_{12}r_2}{\alpha_{11}\alpha_{22} - \alpha_{21}\alpha_{12}} \right) \tag{7.8}$$

$$N_2{}^* = \left(\frac{\alpha_{21}r_1 - \alpha_{11}r_2}{\alpha_{11}\alpha_{22} - \alpha_{21}\alpha_{12}} \right) \tag{7.9}$$

Matrix equation 7.7 can be generalized for any number of species as:

$$-\mathbf{r} = \mathbf{AN}^*$$

where $\mathbf{A}$ is the square matrix of interaction coefficients and $\mathbf{r}$ and $\mathbf{N}^*$ are column matrices of intrinsic rates of change and equilibrium densities respectively. Equilibrium values can then be found by matrix algebra as for equation 7.7 (equivalent to the solution of s simultaneous equations):

$$-\mathbf{r}\mathbf{A}^{-1} = \mathbf{N}^*$$

To determine the community matrix we need to linearize the population growth equation (7.1) at equilibrium. This is achieved with a Taylor expansion or series. A series is defined in mathematics as the sum of a sequence of numbers. We have seen how sequences may arise in ecological processes in Chapter 1 (Fibonacci sequence). Various functions such as e^x or $\sin(x)$ can be expressed as a series. Using a Taylor series a function $f(x+h)$ can be expressed as:

$$f(x+h) = f(x) + hf'(x) + h^2/2! f''(x) + h^3/3! f'''(x) + \ldots$$

where x and h are both variables. $f'(x)$ means the derivative of the function evaluated at x whereas f'' is the second derivative. If x represents the equilibrium density of a population which is described by a nonlinear function then when h is small (a perturbation from the equilibrium) the function near the equilibrium can be expressed according to the Taylor series as:

$$f(x+h) = f(x) + hf'(x) \tag{7.10}$$

That is, ignoring terms with h^2 and higher because h is relatively small. Equation 7.10 describes the linear tangent at equilibrium. In ecological communities we may be dealing with the abundances of many species, all of which have an equilibrium. In this case linearized dynamics are represented as partial derivatives in the Taylor series. Partial differentiation is a method of determining change in one variable while one or more other variables are held constant. Partial derivatives are indicated by ∂. Returning to equation 7.1 and writing it in general terms as:

$$dN_i/dt = F_i(N)$$

the Taylor expansion around the equilibrium is:

$$\frac{dN_i}{dt} = F_i(N^*) + \sum_{j=1}^{k} n_j \left[\frac{\partial\left(\frac{dN_i}{dt}\right)}{\partial N_j} \right]_{N^*} + \text{second- and higher-order terms} \tag{7.11}$$

where n_j is a small perturbation from equilibrium. $F_i(N^*)$ is 0 and second- and higher-order terms can be ignored. The partial derivatives $\partial(dN_i/dt)/\partial N_j$ at equilibrium are $\alpha_{ij}N_i^*$; that is, the interaction coefficient multiplied by the equilibrium population density of the ith species. You could check this for the two-species example. For example, to find $\partial(dN_1/dt)/\partial N_1$ at N^*:

$$\frac{\partial\left(\frac{dN_1}{dt}\right)}{\partial N_1} = r_1 + 2\alpha_{11}N_1^* + \alpha_{12}N_2^* \tag{7.12}$$

Substitute for N_1^* and N_2^* to give:

$$r_1 + \frac{2\alpha_{11}(-\alpha_{22}r_1 + \alpha_{12}r_2)}{\alpha_{11}\alpha_{22} - \alpha_{21}\alpha_{12}} + \frac{\alpha_{12}(\alpha_{21}r_1 - \alpha_{11}r_2)}{\alpha_{11}\alpha_{22} - \alpha_{21}\alpha_{12}}$$

This reduces to:

$$\frac{\alpha_{11}(-r_1\alpha_{22} + \alpha_{12}r_2)}{\alpha_{11}\alpha_{22} - \alpha_{21}\alpha_{12}}$$

which is $\alpha_{11}N_1{}^*$.

The full community matrix for the two-species example is:

$$\begin{pmatrix} \alpha_{11}N_1{}^* & \alpha_{12}N_1{}^* \\ \alpha_{21}N_2{}^* & \alpha_{22}N_1{}^* \end{pmatrix}$$

The stability of the community is found by determining the eigenvalue(s) of the community matrix. This tells us about the growth of a perturbation (n_j) from equilibrium. If the sign of the largest eigenvalue of the community matrix is negative then the community is stable; that is, the perturbations reduce back in size towards the equilibrium. A positive value indicates growth of the perturbation away from the equilibrium. Therefore we can see why the community matrix is sometimes referred to as the stability matrix. The magnitude of the dominant eigenvalue determines the return time of the community (Pimm & Lawton 1977), which measures the time taken for a perturbation to decay to $1/e$ of its initial value.

To conclude this section we link up the graphical interpretation of stability of the logistic equation with the analytical method of the community matrix, following May (1973a) and Pimm (1982). Recall that there are two equilibria with the logistic equation ($N^* = 0$ and $N^* = K$). To examine the stability of those equilibria in Chapter 6 we used a graphical method to examine perturbations (displacements) from equilibrium and asked whether those displacements will become larger with time. If the perturbations do become larger then the equilibrium is locally unstable. From the community matrix analysis we expect that the stable equilibrium of the logistic model is given by a negative slope of dN/dt with respect to N at equilibrium; that is, the single 'eigenvalue' is negative. This is indeed the case (Fig. 7.4).

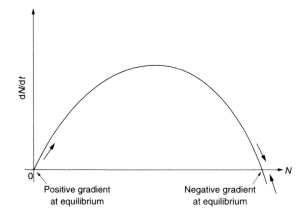

Fig. 7.4 Logistic curve showing unstable and stable equilibria. Note the gradient of the curves at $N^* = 0$ and $N^* = K$.

7.3 Estimations of community stability and structure in the field

Seifert and Seifert (1976) provided one of the earliest field tests of the community matrix using insects in the water-filled bracts of *Heliconia* flowers in Central America (Fig. 7.5). The insects included larvae of chrysomelid beetles and syrphid flies, all of which were potential competitors.

Seifert and Seifert combined experimental manipulations with a multiple regression method which allowed them to estimate the magnitude and signs of the interaction coefficients from a generalized Lotka–Volterra model. This meant that statistical significance could be attached to each of the coefficients. The experiment involved emergent buds of *Heliconia* being enclosed in plastic bags to restrict immigration and oviposition. After a certain amount of growth, water was added and varying numbers of four species of insects were introduced. Following this the per-capita change in numbers with time was determined using a linearized version of equation 7.1, calculated as the change from initial density divided by the number of days over which the change took place. The initial densities of each species were used as the explanatory

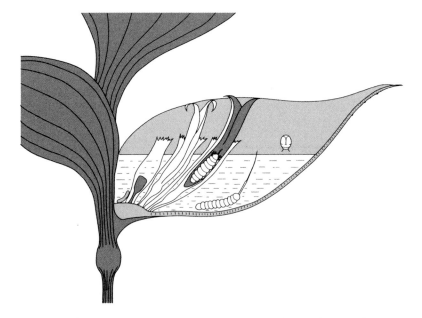

Fig. 7.5 Stylized view of *Heliconia wagneriana* showing the dissected bract with common insect inhabitants. The species include *Gillisius* located on the dissected bract just above the water, *Quichuana* located at the base of the flower below *Gillisius*, *Copestylum* located just inside the flower and *Beebeomyia* located at the base of the seed. From Seifert and Seifert (1976).

variables to calculate the partial regression coefficients of the per capita rates of change against all species; that is, this gave r and α_{ij}. (Note that the rates of change were not estimated from equilibrium as assumed by the community matrix.) A negative value of the regression coefficient indicated competition while a positive value indicated mutualism (the possibility of predation was ignored given the choice of insect species). From Table 7.2 we see that nine of the inter-specific interactions were not significant and therefore were set to zero. Of the significant ones, two were negative (competitors) and one was positive (mutualism).

The equilibrium densities estimated from the model by $N_i^* = A^{-1}r_i$ are shown in Table 7.3 compared with those observed in the field. The fact that there are two negative (unrealistic) densities for $H.\ wagneriana$ suggests that the observed mean densities either are not equilibrium densities or are results of processes not dependent on species interactions, or that the model is inappropriate.

The eigenvalues of the community matrix were determined to examine the stability of the community. The four values were -0.0221, 0.052, -0.042 and -0.239. The positive eigenvalue indicated an unstable community. Seifert and Seifert's conclusion was that $H.\ wagneriana$ insect communities were indeed unstable and that migration, oviposition and local extinction processes may be important in structuring these communities. In other words it is probably not correct to model these communities in isolation. The effects of migration and local extinction are the subject of Chapter 8.

Table 7.2 Interaction matrix for *Heliconia wagneriana*. Non-significant coefficients are set to zero (Seifert & Seifert 1976).

	Quichuana	Gillisius	Copestylum	Beebeomyia
Quichuana	0.001	0	−0.018	0.027
Gillisius	0	−0.003	0	0
Copestylum	0	0	−0.005	0
Beebeomyia	0	−0.005	0	−0.033

Table 7.3 Equilibrium densities predicted from the model compared with mean densities observed in the field (Seifert & Seifert 1976).

	Mean densities in unmanipulated examples	Estimated species equilibrium densities
Quichuana	51.00	−112
Gillisius	7.56	−23.2
Copestylum	8.78	4.09
Beebeomyia	6.67	10.62

Subsequent studies of the community matrix have covered a wide range of species. Wilson and Roxburgh (1992) provided examples of the application of the community matrix to plant species mixtures. They predicted that initially unstable six-species mixtures will, by selective deletion (following Tregonning & Roberts 1979), drop down to stable four-species mixtures. A study of the persistence of chironomid communities in the River Danube demonstrated differences in return times of perturbed communities at different sites (Schmid 1992). An analysis of local and global stability in six small mammal communities showed that all the community matrices were locally and globally stable, due to a reduction in connectance with increasing number of species (Hallett 1991).

The above examples show that it is possible to parameterize community matrix models using field data (with or without manipulations) and make testable predictions about stability, structure and return times after perturbation. Such predictions can be related to species richness and connectance. However, we need to be cautious as analysis of the community matrix is in the neighbourhood of an assumed equilibrium. For many applications we are likely to be interested in communities away from equilibrium or where non-equilibrium processes such as physical disturbance or pollution may be important. Local extinction and colonization processes may also mean that equilibrium has to be judged at larger spatial scales (Chapter 8).

Spatial models

8.1 Spatial dynamics of host–parasitoid systems

In Chapter 6 we noted that Nicholson and Bailey had anticipated that spatial structure, in particular migration and local extinction, would lead to persistence in their model of host–parasitoid interactions. Comins et al. (1992) used the Nicholson–Bailey model to explore the possible outcomes of spatial dynamics. In this study it was assumed that the host and parasitoid were distributed among a square grid of square cells or patches of width n. These types of model systems, known collectively as cellular automata, have been widely used in both plant and animal studies to address a variety of issues including ecological stability and effects of invasive species (e.g. Crawley & May 1987, Silvertown et al. 1992, Colasanti & Grime 1993, Huang et al. 2008; see Wolfram 1984 for a mathematical overview).

Comins et al. (1992) had two phases of dynamics: reproduction/parasitism and dispersal. The former was modelled using the Nicholson–Bailey model. The latter had the following rules.

1 A fraction of the hosts and parasitoids leave the patch (grid cell) and the remainder stay to reproduce in their patch.
2 The fraction dispersing is equally divided between the eight neighbouring patches. There is only one movement per generation. Longer-range dispersal is excluded.
3 There are reflective boundary conditions in which dispersing individuals are prevented from crossing the boundary and remain in the edge patch. Thus there is an explicit edge effect in this model, in contrast to some other cellular automata models.

In small arenas of less than 10 cells by 10 cells extinction of host and parasitoid occurred within a few hundred generations of the simulations, underlining the inherent instability of the Nicholson–Bailey model. However, when the arena size was increased to between 15 and 30 patches, three general types of spatial dynamics were found which Comins et al. described as spirals, spatial chaos and crystal lattices (Fig. 8.1a). The key feature of the three dynamic types is that they permit long-term persistence of the host and parasitoid within a relatively narrow range of population densities (Fig. 8.1b), as

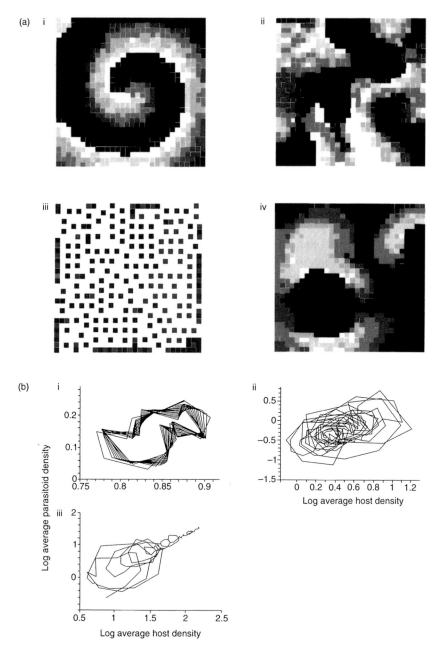

Fig. 8.1 (a) Spatial dynamics of model host and parasitoid. Population density at one point in time from simulations after many generations of the dispersal model of Comins et al. (1992) using an arena width of 30 cells and with Nicholson–Bailey local dynamics. Different levels of shading represent different densities of hosts and parasitoids. Black squares represent empty patches. (i) Spirals, (ii) spatial chaos, (iii) crystalline structures. Case (iv) is a similar diagram obtained with Lotka–Volterra local dynamics which exhibits highly variable spirals. (b) Phase plane showing changes of host and parasitoid over time with the same corresponding parameters as in ai, aii and aiii.

did the incorporation of host density dependence or aggregation of parasit-oids. Similar results were also found with the oscillatory unstable discrete version of the Lotka–Volterra model (Fig. 8.1a, panel iv). Other studies have explored the role of spatial dynamics of predators and prey as a contribution to the stability of their temporal dynamics. For example, McCauley et al. (1993) used an individual-based model to determine the relative importance of predator and prey mobility on stability.

In conclusion, a spatially explicit model can produce long-term population persistence in contrast to an unstable local population model. Mark–release–recapture data collected in the field are shedding light on the short- and long-distance dispersal capabilities of hosts and parasitoids. Jones et al. (1996), in a study of the movements of a tephritid fly (which feeds on thistle seed heads) and its parasitoids showed frequent movements across a patch of thistles of about 50 m × 50 m. The parasitoids moved further than the hosts within the patch. Longer-distance dispersal in similar organisms has been demonstrated by Dempster et al. (1995) using rubidium and other chloride salts in plants to mark herbivores and parasitoids. This work demonstrated that distances of up to 800 m are not a barrier to colonization.

In the next section we develop the theme of spatial models, focusing on analytical techniques rather than on the results of cellular automata simulations.

8.2 Metapopulation models

8.2.1 Introduction to the metapopulation concept

A metapopulation is defined as a set of local populations linked by dispersal. This could be described and modelled by cellular automata but we will focus on results arising from analytical considerations. In the original model of Levins (1969, 1970) it was assumed that all local populations were of equal size and that a local population could either become extinct or reach carrying capacity instantaneously following colonization. Therefore only two states of local population were envisaged: full (carrying capacity) or empty (extinct).

In reality, the definition of a local population, and therefore a metapopula-tion, is very difficult. Hanski and Gilpin (1991) defined a local population as a 'set of individuals [of the same species] which all interact with each other with a high probability'. But how high is that probability? Furthermore, 'local' may be different for different interactions. For example, two plants may show intraspecific competition over a scale of a few centimetres but be reproductively linked by pollination over hundreds of metres. It is also very difficult to say over what distance colonization of new areas, and therefore the 'birth' of new local populations, may occur. Typically the frequency of movements of propagules such as seeds over short distances is known, but longer-distance movement is poorly known, partly because it may be a rare event and partly because it is difficult to record. This excludes species which show seasonal and predictable long-distance migration.

Even when local populations can be identified, the pure Levins model of local populations with equal carrying capacity is unusual. More realistically, it is reasonable to envisage a spectrum of possibilities from mainland/island or core/satellite to pure Levins populations (Fig. 8.2). These and other possi-

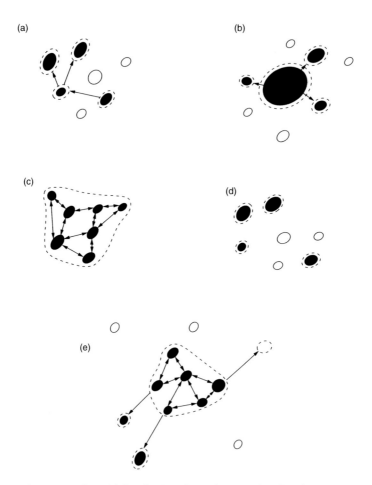

Fig. 8.2 Various types of spatial distribution of populations. Closed ovals represent occupied habitat patches and open ovals represent vacant habitat patches. Dashed lines indicate the boundaries of populations. Arrows indicate migration and colonization. (a) Levins metapopulation. (b) Core/satellite metapopulation (Boorman & Levitt 1973). (c) Patchy population. (d) Non-equilibrium metapopulation (differs from (a) in that there is no recolonization). (e) An intermediate case that combines (b) and (c) (Harrison 1991).

bilities have been discussed by Harrison (1991), who considered the rarity of true Levins metapopulations in the field, and by Hanski and Gyllenberg (1993), who showed how to model both mainland/island and pure Levins with related equations. Hanski (1999) provides an overview of the whole subject.

Various processes will promote something close to a metapopulation structure in the field or at least create conditions under which local extinction and colonization are integral features of the population dynamics:

- gap creation or other disturbance generating new recruitment habitat (includes habitat fragmentation);
- a mosaic of successional habitats where, for example, an annual plant must move from one transient early successional habitat to another (may be a function of the previous feature);
- sedentary and localized resources such as plants, dung or decaying logs, all of which may be colonized by various insects, fungi and other organisms. The resources may have a short colonization period allowing a maximum of one or a few generations of the attacking organism, thereby necessitating dispersal to other similar resources.

8.2.2 The metapopulation model of Levins

Despite the problems of finding Levins metapopulations in the field it is instructive to consider its dynamics before proceeding to more complex models. Levins (1969, 1970) was interested in the number of islands or island-like habitats occupied by a species. Later Levins and Culver (1971) modified the model to investigate the effect of competition on migration and extinction rates. Levins began by considering the number of local populations (N), the total number of sites (T), an extinction rate (e) and a migration rate (m'). The rate of change of N with time (t) could then be expressed as a differential equation:

$$\mathrm{d}N/\mathrm{d}t = m'N(T - N) - eN$$

This equation was simplified by using $p = N/T$ where p represents the fraction of habitat patches occupied by a species and replacing $m'T$ by m. The rate of change in the fraction of habitat patches occupied by a species, $\mathrm{d}p/\mathrm{d}t$ – that is, the rate of change in the proportion of local populations (p) at a given time – was now described by:

$$\mathrm{d}p/\mathrm{d}t = mp(1 - p) - ep \tag{8.1}$$

where m defines the colonization rate of local populations and e the extinction rate of local populations. Therefore ep represents loss (or extinction) of local populations from metapopulations. The birth rate of local populations is represented by $mp(1 - p)$. The reason why p is multiplied by $1 - p$ can be conceptualized as a local neighbourhood problem. If there is one occupied patch surrounded by eight empty patches then the probability of colonization of any one empty patch is likely to be less than if there was one empty patch surrounded by eight occupied patches. Thus in determining the colonization probability the density of occupied patches and unoccupied patches needs to be combined, for example by multiplying them. In reality, the colonization (m) and extinction (e) parameters are likely to be complex functions of a set of variables. For example, m involves finding a new site, which depends on propagule dispersal (in turn dependent on the taxon and habitat under scrutiny and perhaps wind or water current speed or abundance of animal dispersers), the spatial distribution of occupied and unoccupied sites, initial establishment of propagules and subsequent population growth.

Now let us consider the dynamics of the system described by equation 8.1. What are the conditions for metapopulation increase, no change or decline? No change in p is given by $dp/dt = 0$:

$$0 = mp(1 - p) - ep$$

which gives

$$p = 1 - e/m$$

Increase in p will occur if $dp/dt > 0$ and therefore $p < 1 - e/m$. In considering these results we need to think about the original formulation of the metapopulation concept. If extinction and colonization rates (death and birth) are balanced ($e = m$) there should be no change in metapopulation size. Similarly, if the extinction rate (e) is greater than the colonization rate (m) then the metapopulation size should decrease (and vice versa). These requirements are only partly supported by the manipulation of equation 8.1. The problem is that when $dp/dt = 0$, if $e = m$ we are left with the result that $p = 1 - 1 = 0$. Thus, when extinction balances colonization there are no local populations. If colonization is greater than extinction then e/m is less than 1 but greater than 0 and therefore $1 - e/m$ lies between 1 and 0. Therefore a steady value of p occurs when colonization exceeds extinction. Hanski (1991) considers various refinements and developments of the basic Levins model. Despite the drawbacks of equation 8.1 we will see in the next section how it can be used to explain limits to species range and how more complex models can give similar predictions.

8.2.3 The Carter and Prince model of geographic range

The plant metapopulation model of Carter and Prince (1981, 1988) linked the ideas of Levins with models of infectious disease to provide an explanation for the geographical range limits of plant species. In particular, they challenged the view that distribution was determined solely by correlation to climate variables; for example, that the northerly distribution limit of plant species in Britain was determined by physiological intolerance of cold winters (see examples in Carter & Prince 1988). Carter and Prince used a differential equation to describe a strategic model of plant distribution:

$$dy/dt = bxy - cy \qquad (8.2)$$

where x is the number of susceptible sites (sites available to be colonized), y is the number of infective sites (occupied sites from which seed is produced and dispersed), b is the infection rate and c is the removal rate. b and c are essentially local population birth (colonization) and death (extinction) rates and therefore equivalent to m and e in the Levins equation (8.1). Similarly, x and y are related to $1 - p$ and p where p is the proportion of local populations which are occupied and potentially 'infective' and $1 - p$ is the proportion of vacant and therefore susceptible sites.

These comparisons show that any conclusions from the Carter and Prince model are relevant to metapopulations in general as defined by Levins. The important conclusion of Carter and Prince was that, along a climatic gradient, a very small change in, for example, temperature might tip the balance from metapopulation persistence to metapopulation extinction. In Carter and Prince's words: 'a climatic factor might lead to distribution limits that are abrupt relative to the gradient in the factor, even though the physiological responses elicited might appear too small to explain such limits.' Thus climate and physiological factors are still important but their effects are amplified and made nonlinear by the threshold properties of equation 8.2.

The results of these simple models are supported by the conclusions from more complex models. For example, the model of Herben, Rydin and Soderstrom (1991) examined the dynamics of the moss *Orthodontium lineare* which occurs on temporary substrates such as rotting wood. The model addressed not only the metapopulation structure of the moss but also the fact that the species was spreading throughout western and central Europe. The model included deterministic increase on occupied logs with a carrying capacity and the assumptions that dispersal by spores was in proportion to local population size and that spore dispersal distance declined exponentially from an occupied log. The results of this model suggested, like the simpler models, that there is a threshold for metapopulation persistence. In this case the percentage of logs occupied was a nonlinear function of probability of local population establishment (p_{est}; Fig. 8.3).

At a p_{est} value of about 0.0002 the model predicted a sudden increase in the percentage of occupied sites. So here was a threshold value above which metapopulation persistence was likely to be high. If p_{est} is a function of climate then this would produce exactly the type of sharp break in species range predicted by Carter and Prince. It seems that such thresholds may be generated in a variety of ways. Below we consider how changes in gap frequency in grassland can generate a threshold for plant population abundance and how diffusion processes can lead to thresholds.

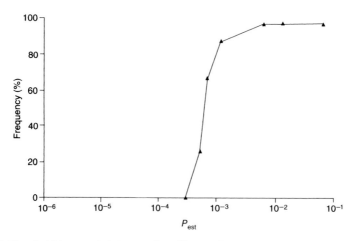

Fig. 8.3 Threshold in occupied sites produced by small increments in probability of establishment (model of Herben et al. 1991).

An alternative formulation for the metapopulation equation 8.1 is to use the logistic equation to model metapopulation dynamics and consider again the possibility of a threshold determining the edge of a species range. One key feature of the model needs to be retained; that is, that there is some interaction between the densities of infectives (occupied patches) and susceptibles (empty patches) in determining colonization rates. This interaction is represented by $p(1 - p)$ in equation 8.1 and xy in equation 8.2. This can be taken further by considering the relationship between the relative or net colonization rate (m/e or $m - e$) and the density of susceptibles with respect to infectives (S/I). In the absence of any effect of S/I the change in infectives (dI/dt) can be described as the net colonization rate multiplied by the number of susceptible patches:

$$dI/dt = (m - e)S \qquad (8.3)$$

The relative colonization rate $m - e$ will be expected to vary with S and I. If S/I is high then $m - e$ should be low. If S/I is low then we expect $m - e$ to be close to its maximum value. There is a clear analogy with the logistic equation. $m - e$ can be replaced by a value r (births minus deaths) and S/I by S'. The simplest reduction of r is linear with respect to S' which is described by $1 - S'/K$. The resultant equation is:

$$dI/dt = rS(1 - S'/K) \qquad (8.4)$$

Note that, in contrast to the logistic equation, the rate of change variable (I) on the left-hand side is not the same as the variable on the right-hand side. At equilibrium, $dI/dt = 0$ so $rS = 0$, which can be interpreted as $r = 0$ ($e = m$), and/or $S = 0$ (no remaining susceptible sites) or $1 - S'/K = 0$ and therefore

$S' = K$ (as expected from the logistic equation). If $S' > K$ then $1 - S'/K$ is negative and so dI/dt is negative. Therefore I decreases and consequently S/I continues to increase. Thus K is a threshold condition. If S' is too high (above K) then the metapopulation cannot persist and the number of infectives declines (and so the proportion of susceptibles, S', increases).

8.3 Gap models and plant population thresholds

Plant population studies in the 1980s (Crawley & May 1987, Klinkhamer & De Jong 1989, Silvertown & Smith 1989) described apparent thresholds for plant population persistence determined by gap density in grassland (Fig. 8.4).

These studies showed that for short-lived herbaceous plants such as *Cirsium vulgare* small changes in gap density due to, for example, disturbance by

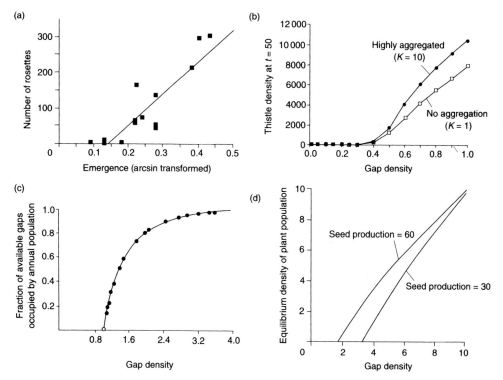

Fig. 8.4 Threshold for plant population persistence with a change in gap density.
(a) Field data showing the relationship between *Cirsium vulgare* rosette numbers and the probability of emergence of seeds sown (Silvertown & Smith 1989); (b–d) Predictions from models of (b) Silvertown and Smith (1989), (c) Crawley and May (1987) and (d) Klinkhamer and De Jong (1989).

grazing animals or absence of perennials, resulted in large changes in the density of plants recruiting solely by seed. A simple model below explains this threshold and illustrates how it can be related to the simulation models of Crawley and May (1987) and Silvertown and Smith (1989) and the analytical model of Klinkhamer and De Jong (1989). The threshold arises directly from a spatially explicit model in which the seed are distributed in a particular way across a set of gaps.

Assume a field of area A is covered with n gaps of equal size (g); the area of gaps $= ng$ and the fraction of field covered by gaps (f) is:

$$f = ng/A$$

Now consider the proportion of gaps receiving one or more seeds, as only these seeds may be expected to germinate. Assume that a maximum of one seed can germinate per gap. To estimate the proportion of gaps receiving one or more seeds we could assume that the seeds are distributed according to the Poisson distribution (Chapter 3). The proportion of gaps containing no seeds is e^{-m} where m is the mean number of seeds per gap. Therefore the proportion of gaps containing one or more seeds is $1 - e^{-m}$. This was assumed by Crawley and May (1987) and Klinkhamer and De Jong (1989). Similarly, we could assume a negative binomial distribution with no aggregation which was used in Silvertown and Smith's simulation (Fig. 8.4b; the degree of aggregation did not affect the outcome) and which we will use below. The effect of distributing seeds between gaps in this way, with a maximum of one survivor per gap, is to introduce density dependence into the model.

With the negative binomial and no aggregation of seed the proportion of gaps containing one or more seeds is $m/m + 1$ where m, the mean number of seeds per gap, is given by the total number of seeds (s) multiplied by the fraction falling into gaps divided by the number of gaps (n). We will assume that the fraction of seed falling into gaps is equivalent to the fraction of ground covered by gaps (f). Therefore

$$m = sf/n \qquad (8.5)$$

Now let us incorporate these details into a model of biennial population dynamics. The number of first-year rosettes (R) in year $t + 1$ is given by the proportion of gaps which contain one or more seeds $(m/m + 1)$ multiplied by the number of gaps (n) and the probability of survival from seed to rosette (p_1):

$$R_{t+1} = mnp_1/m + 1 \qquad (8.6)$$

Substitute the right-hand side of equation 8.5 for m in the numerator in equation 8.6 and cancel n:

$$R_{t+1} = sfp_1/m + 1 \qquad (8.7)$$

Now assume that the first-year rosettes survive with probability p_2 to become flowering plants (F) in the next year:

$$F_{t+2} = p_2 R_{t+1}$$

and that, in the same year, each flowering plant produces an average of q seeds. Therefore, the total number of seeds (s) in year $t + 2$ is related to the number of rosettes in the previous year ($t + 1$):

$$s_{t+2} = q p_2 R_{t+1} \qquad (8.8)$$

Substitute the right-hand side of equation 8.7 into equation 8.8:

$$s_{t+2} = (s_t f p_1 / m + 1) q p_2$$

Let $q p_1 p_2 = \lambda$ to give:

$$s_{t+2} = s_t f \lambda / (m + 1) \qquad (8.9)$$

Set equation 8.9 at equilibrium (and cancel s^*) and rearrange to make m the subject of the equation:

$$m = f \lambda - 1$$

Now substitute for m from equation 8.5:

$$s^* f / n = f \lambda - 1$$

$$s^* = n \lambda - (n / f)$$

We know that $f = ng / A$ and therefore $n / f = A / g$. A / g represents the ratio of the size of the field to the size of the gap and can be replaced by α:

$$s^* = n \lambda - \alpha \qquad (8.10)$$

The relationship between the equilibrium number of seeds (s^*) and the gap density (n) is shown in Fig. 8.5, illustrating the threshold effect of gap density. If n_T is the threshold gap density, then when $n < n_T$ the result is that $s^* = 0$ (s^* can theoretically also be negative but this is not ecologically realistic). The analytical results of equation 8.10 therefore support the results of the simulations of Silvertown and Smith (1989). The model of Klinkhamer and de Jong (1989) reached similar conclusions showing that if $dsc < 1$ (where d is the density of gaps, s is the seed production and c is the gap area) the population would become extinct and if $dsc > 1$ then a population equilibrium would exist. This may begin to sound familiar. dsc is acting like a finite rate of increase for a gap-dependent grassland plant. The same is true of $n\lambda$ in equation 8.10. The threshold result is therefore as predicted by the discrete-time density-independent and density-dependent models; that is, the requirement for $\lambda \geq 1$ for a population not to decline towards zero. The results in this section are applicable beyond plant populations in grassland where recruitment into gaps is important;

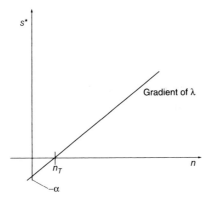

Fig. 8.5 Threshold in s^* predicted from equation 8.10.

for example, seedlings of tree species into forest gaps or planktonic larvae of barnacles into gaps in the rocky shore.

8.4 Diffusion processes

Many population models ignore the details of the dispersal phase. Too little is usually known about dispersal to model it in anything other than the simplest way (but see the work of Pacala on neighbourhood models for an exception to this, e.g. Pacala & Silander 1990). One simple approximation which leads to interesting results is to assume that the individuals diffuse out from a source population. This may describe the expansion of an invading species across suitable habitat or the movement of individuals between local populations across uncolonizable habitat. Diffusion is a random and continuous process with each particle or individual going on a random walk from its source position. Although the concept is straightforward, diffusion models are complex because they require a method of summarizing all the random movements at each point in time. Applications of spatial diffusion models in ecology include the work of Morris (1993) on pollen dispersal and insect movement and marine ecologists studying the movement of algae in water bodies (see below). Segel (1984) gives an introduction to diffusion models of bacteria movement.

Diffusion can be described by partial differential equations, or PDEs. These are needed because movement and/or abundance of individuals is dependent on two variables: spatial position and time. Maynard Smith (1968) provides an accessible introduction to partial differentiation applied to biological problems. He describes the diffusion of a substance along a tube. The change in concentration (x) with time (t) is related to the change in concentration with distance (s) according to the following PDE:

$$\frac{\partial x}{\partial t} = \mu \frac{\partial x^2}{\partial s^2}$$

(8.11)

where μ is a constant. This equation is well known in the mathematical literature as the one-dimensional heat equation. An ecological use of this equation is the dispersal of individuals along a linear route, such as plants dispersing along a road side. In this case the constant might combine finite rate of change and mean dispersal distance. A major problem is that only linear PDEs have analytical solutions. Despite this it is worth considering the results of PDE analyses of diffusion processes in ecology as they have produced results which support the results of boundaries and thresholds above. Also, as Maynard Smith observed (and this applies to many mathematical results), 'a familiarity with the notation enables one to follow other people's arguments, even if one could not have developed the argument oneself'.

A relatively simple model was used by Kierstead and Slobodkin (1953) who related the size of a plankton patch to the degree of turbulent diffusion. Kierstead and Slobodkin wanted to determine whether or not there is a minimum water mass size below which no increase in phytoplankton concentration is possible. They determined the existence of a threshold condition for a patch of algae to cause a red tide. Steele (1974) also looked at diffusion in marine systems, again asking what causes patchiness of algae in the sea. He considered that 'lateral turbulent diffusion of the water is a dominant physical process and can be expressed in a mathematical formulation'. Steele compared this physical process with the ecological process of herbivory using a pair of PDEs and showed that if the system without turbulence is unstable then diffusion under certain conditions could stabilize the system and, above a critical value, the system would destabilize again. Steele concluded that 'if an ecosystem is basically unstable when considered without diffusion processes then diffusion can remove the instability at smaller scales but not larger ones'.

Spatially explicit models continue to be one of the most exciting areas of ecological research. Increasingly, a spatial dimension is being combined into temporal models with interplay between the different concepts covered in this book. Spatial models reveal how independent lines of enquiry can lead to similar ecological principles; for example, the production of thresholds. They also have important implications for the relationship between theory and application. For example, metapopulation models have demonstrated the clear requirement for detailed field study of colonization and extinction rates and associated parameters such as fragmentation of habitat (Hanski's analysis of metapopulations is an exemplar of combining theory and field work). Spatial models caution us against simplistic interpretations of large spatial scale anthropogenic effects such as global climate change. The preponderance of threshold possibilities shows that we must expect a nonlinear response to climate change. Species will not simply shift their range in linear procession;

there will be extinctions and outbreaks as predators and prey uncouple, metapopulation parameters are tweaked and finite rates of increase shift around their thresholds. These applications alone should encourage us to take greater interest in the methods and results of mathematical models. With such applications in mind, I should remark on a major omission from this book: ecosystem models. While many of the principles covered here, such as stability and spatial methods, apply to ecosystem modelling, there is no doubting the need for further understanding of ecosystem models, especially those aimed at understanding the biogeochemical properties and dynamics of the whole Earth. Fortunately there is no shortage of texts and review articles covering the subject.

References

Alfonso-Corrado, C., Clark-Tapia, R. and Mendoza, A. (2007) Demography and management of two clonal oaks: *Quercus eduardii* and *Q. potosina* (Fagaceae) in central Mexico. *Forest Ecology and Management* 251, 129–41.

Allee, W.C., Emerson, A.E., Park, O., Park, T. and Schmidt, K.P. (1949) *Principles of Animal Ecology*. W.B. Saunders, Philadelphia, PA.

Atkinson, W.D. and Shorrocks, B. (1981) Competition on a divided and ephemeral resource: a simulation model. *Journal of Animal Ecology* 50, 461–71.

Baltensweiler, W. (1993) Why the larch bud-moth cycle collapsed in the subalpine larch-cembran pine forests in the year 1990 for the first time since 1850. *Oecologia* 94, 62–6.

Bambach, R.K. (2006) Phanerozoic biodiversity mass extinctions. *Annual Review of Earth and Planetary Science* 34, 127–55.

Beddington, J.R. (1979) Harvesting and population dynamics. In: *Population Dynamics* (R.M. Anderson, B.D. Turner and L.R. Taylor, eds), pp. 307–20. Blackwell Science, Oxford.

Beddington, J.R., Free, C.A. and Lawton, J.H. (1975) Dynamic complexity in predator-prey models framed in difference equations. *Nature* 255, 58–60.

Bernardelli, H. (1941) Population waves. *Journal of the Burma Research Society* 31, 1–18.

Berryman, A.A. (1992) On choosing models for describing and analyzing ecological time series. *Ecology* 73, 694–8.

Beverton, R.J.H. and Holt, S.J. (1957) On the dynamics of exploited fish populations. *Fishery Investigations, London Series* 2(19).

Boorman, S.A. and Levitt, P.R. (1973) Group selection on boundary of a stable population. *Theoretical Population Biology* 4, 85–128.

Broekhuizen, N., Evans, H.F. and Hassell, M.P. (1993) Site characteristics and the population dynamics of the pine looper moth. *Journal of Animal Ecology* 62, 511–18.

Buckley, Y.M., Hinz, H.L., Matthies, D. and Rees, M. (2001) Interactions between density-dependent processes, population dynamics and control of an invasive plant species, *Tripleurospermum perforatum* (scentless chamomile). *Ecology Letters* 4, 551–8.

Callaway, R.M. and Davis, F.W. (1993) Vegetation dynamics, fire and the physical environment in coastal central California. *Ecology* 74, 1567–78.

Carter, R.N. and Prince, S.D. (1981) Epidemic models used to explain biogeographical distribution limits. *Nature* 293, 644–5.

Carter, R.N. and Prince, S.D.(1988) Distribution limits from a demographic viewpoint. In: *Plant Population Ecology* (A.J. Davy, M.J. Hutchings and A.R. Watkinson, eds), pp. 165–84. Blackwell Science, Oxford.

Caswell, H. (2000a) *Matrix Population Models: Construction, Analysis and Interpretation*. Sinauer Associates, Sunderland, MA.

Caswell, H. (2000b) Prospective and retrospective perturbation analyses: their roles in conservation biology. *Ecology* 81, 619–27.

Colasanti, R.L. and Grime, J.P. (1993) Resource dynamics and vegetation processes: a deterministic model using two dimensional cellular automata. *Functional Ecology* 7, 169–76.

Comins, H.N., Hassell, M.P. and May, R.M. (1992) The spatial dynamics of host-parasitoid systems. *Journal of Animal Ecology* 61, 735–48.

Cook, R.M., Sinclair, A. and Stefansson, G. (1997) Potential collapse of North Sea cod stocks. *Nature* 385, 521–2.

Cornette, J.L. and Lieberman, B.S. (2004) Random walks in the history of life. *Proceedings of the National Academy of Sciences USA* 101, 187–91.

Crawley, M.J. and May, R.M. (1987) Population dynamics and plant community structure: competition between annuals and perennials. *Journal of Theoretical Biology* 125, 475–89.

Crombie, A.C. (1945) On competition between different species of granivorous insects. *Proceedings of the Royal Society of London Series B Biological Sciences* 132, 362–95.

Crombie, A.C. (1946) Further experiments on insect competition. *Proceedings of the Royal Society of London Series B Biological Sciences* 133, 76–109.

Crombie, A.C. (1947) Interspecific competition. *Journal of Animal Ecology* 16, 44–73.

de Kroon, H., Plaisier, A. and van Groenendael, J. (1987) Density-dependent simulation of the population dynamics of a perennial grassland species, *Hypochaeris radicata*. *Oikos* 50, 3–12.

de Kroon, H., van Groenendael, J. and Ehrlen, J. (2000) Elasticities: a review of methods and model limitations. *Ecology* 81, 607–18.

Dempster, J.P., Atkinson, D.A. and French, M.C. (1995) The spatial dynamics of insects exploiting a patchy food resource. II. Movements between patches. *Oecologia* 104, 354–62.

Dennis, B. and Taper, M.L. (1994) Density dependence in time series observations of natural populations: estimation and testing. *Ecological Monographs* 64, 205–24.

Elliott, J.M. (1994) *Quantitative Ecology and the Brown Trout*. Oxford Series in Ecology and Evolution. Oxford University Press, Oxford.

Ellner, S. and Turchin, P. (1995) Chaos in a noisy world: new methods and evidence from time-series analysis. *American Naturalist* 245, 343–75.

Elton, C.S. (1958) *The Ecology of Invasions by Animals and Plants*. Methuen, London.

Elton, C.S. and Nicholson, M. (1942) The ten-year cycle in numbers of the lynx in Canada. *Journal of Animal Ecology* 11, 215–44.

Enquist, B.J., Brown, J.H. and West, G.B. (1998) Allometric scaling of plant energetics and population density. *Nature* 395, 163–5.

Federico, P. and Canziani, A. (2005) Modeling the population dynamics of capybara *Hydrochaeris hydrochaeris*: a first step towards a management plan. *Ecological Modelling* 186, 111–21.

Fieberg, J. and Ellner, S.P. (2001) Stochastic matrix models for conservation and management: a comparative review of methods. *Ecology Letters* 4, 244–66.

Flores-Moya, A., Fernandez, J.A. and Niell, F.X. (1996) Growth pattern, reproduction and self-thinning in seaweeds. *Journal of Phycology* 32, 767–9.

Foley, P. (1994) Predicting extinction times from environmental stochasticity and carrying capacity. *Conservation Biology* 8, 124–37.

Franco, M. and Silvertown, J. (2004) Comparative demography of plants based upon elasticities of vital rates. *Ecology* 85, 531–8.

Freckleton, R.P., Sutherland, W.J., Watkinson, A.R. and Stephens, P.A. (2008) Modelling the effects of management on population dynamics: some lessons from annual weeds. *Journal of Applied Ecology* 45, 1050–8.

Gardner, M.R. and Ashby, W.R. (1970) Connectance of large dynamic (cybernetic) systems: critical values for stability. *Nature* 228, 784.

Gauci, V., Dise, N. and Fowler, D. (2002) Controls on suppression of methane flux from a peat bog subjected to stimulated acid rain sulphate deposition. *Global Biogeochemical Cycles* 16(1), article 1004.

Gause, G.F. (1932) Experimental studies on the struggle for existence. I. Mixed populations of two species of yeast. *Journal of Experimental Biology* 9, 389–402.

Gause, G.F. (1934) *The Struggle for Existence*. Williams and Wilkins, Baltimore, MD. Reprinted 1964, Hafner, New York.

Gause, G.F. (1935) Experimental demonstration of Volterra's periodic oscillation in the number of animals. *Journal of Experimental Biology* 12, 44–8.

Gillman, M.P. and Crawley, M.J. (1990) A comparative evaluation of models of cinnabar moth dynamics. *Oecologia* 82, 437–45.

Gillman, M.P. and Dodd, M. (2000) Detection of delayed density dependence in an orchid population. *Journal of Ecology* 88, 204–12.

Gillman, M.P., Bullock, J.M., Silvertown, J. and ClearHill, B. (1993) A density dependent model of *Cirsium vulgare* population dynamics using field estimated parameter values. *Oecologia* 96, 282–9.

Gilpin, M.E. (1979) Spiral chaos in a predator-prey model. *American Naturalist* 107, 306–8.

Guckenheimer, J. and Holmes, P. 1983. *Nonlinear Oscillations, Dynamical Systems and Bifurcations of Vector Fields*. Springer-Verlag, New York.

Hallett, J.G. (1991) The structure and stability of small mammal faunas. *Oecologia* 88, 383–93.

Hanski, I. (1991) Single-species metapopulation dynamics: concepts, models and observations. *Biological Journal of the Linnean Society* 42, 17–38.

Hanski, I. (1999) *Metapopulation Ecology*. Oxford University Press, Oxford.

Hanski, I. and Gilpin, M. (1991) Metapopulation dynamics: brief history and conceptual domain. *Biological Journal of the Linnean Society* 42, 3–16.

Hanski, I. and Gyllenberg, M. (1993) Two general metapopulation models and the core-satellite hypothesis. *American Naturalist* 142, 17–41.

Harrison, S. (1991) Local extinction in a metapopulation context: an empirical evaluation. *Biological Journal of the Linnean Society* 42, 73–88.

Hassell, M.P. (1975) Density dependence in single species populations. *Journal of Animal Ecology* 42, 693–726.

Hassell, M.P. (1976) *The Dynamics of Competition and Predation*. Arnold, London.

Hassell, M.P. and Comins, H.N. (1976) Discrete time models for two-species competition. *Theoretical Population Biology* 9, 202–21.

Hassell, M.P. and Godfray, H.C.J. (1992) The population biology of insect parasitoids. In: *Natural Enemies* (M.J. Crawley, ed.), pp. 265–92. Blackwell Science, Oxford.

Hassell, M.P. and May, R.M. (1973) Stability in insect host-parasite models. *Journal of Animal Ecology* 42, 693–726.

Hassell, M.P. and Varley, G.C. (1969) New inductive population model for insect parasites and its bearing on biological control. *Nature* 223, 1133–7.

Hassell, M.P., Lawton, J.H. and May, R.M. (1976) Patterns of dynamical behaviour in single species populations. *Journal of Animal Ecology* 45, 471–86.

Hastings, A. and Powell, T. (1991) Chaos in a three-species food chain. *Ecology* 72, 896–903.

Herben, T., Rydin, H. and Soderstrom, L. (1991) Spore establishment probability and the persistence of the fugitive invading moss, *Orthodontium lineare*: a spatial simulation model. *Oikos* 60, 215–21.

Holling, C.S. (1966) The strategy of building models in complex ecological systems. In: *Systems Analysis in Ecology* (K.E.F. Watt, ed.). Academic Press, New York.

Holling, C.S. (1973) Resilience and stability of ecological systems. *Annual Review of Ecology and Systematics* 4, 1–23.

Holmes, E.E., Sabo, J.L., Viscido S.V. and Fagan, W.F. (2007) A statistical approach to quasi-extinction forecasting. *Ecology Letters* 10, 1182–98.

Holst, N., Rasmussen, I.A. and Bastiaans, L. (2007) Field weed population dynamics: a review of model approaches and applications. *Weed Research* 47, 1–14.

Horn, H.S. (1975) Markovian properties of forest succession. In *Ecology and Evolution of Communities* (M.L. Cody and J.M. Diamond, eds), pp. 196–211. Harvard University Press, Cambridge, MA.

Horn, H.S. (1981) Succession. In: *Theoretical Ecology* (ed. R.M. May), pp. 253–71. Blackwell Science, Oxford.

Huang, H., Zhang, L., Guan, Y. and Wang, D. (2008) A cellular automata model for population expansion of *Spartina alterniflora* at Jiuduansha Shoals, Shanghai, China. *Estuarine, Coastal and Shelf Science* 77, 47–55.

Huffaker, C.B. (1958) Experimental studies on predation. Dispersion factors and predator-prey oscillations. *Hilgardia* 27, 343–83.

Hui, D. and Jackson, R.B. (2006) Geographical and interannual variability in biomass partitioning in grassland ecosystems: a synthesis of field data. *New Phytologist* 169, 85–93.

Hutchinson, G.E. (1948) Circular causal systems in ecology. *Annals of the New York Academy of Science* 50, 221–46.

Hutchinson, G.E. (1978) *An Introduction to Population Ecology.* Yale University Press, New Haven, CT.

Jansen, V.A.A. and Kokkoris, G.D. (2003) Complexity and stability revisited. *Ecology Letters* 6, 498–502.

Jones, T.H., Godfray, H.C.J. and Hassell, M.P. (1996) Relative movement patterns of a tephritid fly and its parasitoid wasps. *Oecologia* 106, 317–24.

Kierstead, H. and Slobodkin, L.B. (1953) The size of water masses containing algal blooms. *Journal of Marine Research* 12, 141–7.

Kingsland, S.E. (1985) *Modeling Nature. Episodes in the History of Population Ecology.* University of Chicago Press, Chicago, IL.

Klinkhamer, P.G.L. and De Jong, T.J. (1989) A deterministic model to study the importance of density dependence for regulation and the outcome of intra-specific competition in populations of sparse plants. *Acta Botanica Neerlandica* 38, 57–65.

Krebs, C.J. (1994) *Ecology.* Harper Collins, New York.

Lafferty, K.D., Hechinger, R.F., Shaw, J.C., Whitney, K.L. and Kuris, A.M. (2006) Food webs and parasites in a salt marsh ecosystem. In: *Disease Ecology: Community Structure and Pathogen Dynamics* (S. Collinge and C. Ray, eds), pp. 119–34. Oxford University Press, Oxford.

Lafferty, K.D. Allesina, S., Arim, M., Briggs, C.J., De Leo, G., Dobson, A.P. et al. (2008) Parasites in food webs: the ultimate missing links. *Ecology Letters* 11, 533–46.

Lefkovitch, L.P. (1965) The study of population growth in organisms grouped by stages. *Biometrics* 21, 1–18.

Lefkovitch, L.P. (1967) A theoretical evaluation of population growth after removing individuals from some age groups. *Bulletin of Entomological Research* 57, 437–45.

Leslie, P.H. (1945) On the uses of matrices in certain population mathematics. *Biometrika* 33, 182–212.

Leslie, P.H. (1948) Some further notes on the use of matrices in population mathematics. *Biometrika* 35, 213–45.

Levins, R. (1966) The strategy of model building in population biology. *American Scientist* 54, 421–31.

Levins, R. (1968) *Evolution in Changing Environments*. Princeton University Press, Princeton, NJ.

Levins, R. (1969) Some demographic and genetic consequences of environmental heterogeneity for biological control. *Bulletin of the Entomological Society of America* 15, 237–40.

Levins, R. (1970) Extinction. In: *Some Mathematical Problems in Biology* (M. Gerstenhaber, ed.), pp. 77–107. Mathematical Society, Providence, RI.

Levins, R. and Culver, D. (1971) Regional coexistence of species and competition between rare species. *Proceedings of the National Academy of Sciences USA* 68, 1246–8.

Lewis, E.G. (1942) On the generation and growth of a population. *Sankhya* 6, 93–6.

Lewontin, R.C. and Cohen, D. (1969) On population growth in a randomly varying environment. *Proceedings of the National Academy of Sciences USA* 62, 1056–60.

Lonsdale, W.M. (1999) Global patterns of plant invasions and the concept of invasibility. *Ecology* 80, 1522–36.

Lotka, A.J. (1925) *Elements of Physical Biology*. Williams and Wilkins, Baltimore, MD. Reprinted 1956, Dover Publications, New York.

Lotka, A.J. (1927) Fluctuations in the abundance of species considered mathematically (with comment by V. Volterra). *Nature* 119, 12–13.

Mace, G.M. and Lande, R. (1991) Assessing extinction threats: towards a re-evaluation of IUCN threatened species categories. *Conservation Biology* 5, 148–57.

Magallon, S. and Sanderson, M.J. (2001) Absolute diversification rates in angiosperm clades. *Evolution* 55, 1762–80.

Manly, B.J. (1990) *Stage-Structured Populations. Sampling, Analysis and Simulation*. Chapman and Hall, London.

Martinez, N.D., Hawkins, B.A., Dawah, H.A. and Feifarek, B.P. (1999) Effects of sampling effort on characterization of food-web structure. *Ecology* 80, 1044–55.

May, R.M. (1972) Will a large complex system be stable? *Nature* 238, 413–14.

May, R.M. (1973a) *Stability and Complexity in Model Ecosystems*. Princeton University Press, Princeton, NJ.

May, R.M. (1973b) On relationships among various types of population model. *American Naturalist* 107, 46–57.

May, R.M. (1976) Simple mathematical models with very complicated dynamics. *Nature* 261, 459–67.

May, R.M. (1978) Host-parasitoid systems in patchy environments, a phenomenological model. *Journal of Animal Ecology* 47, 833–43.

May, R.M. (1981) *Theoretical Ecology. Principles and Applications*. Blackwell Science, Oxford.

May, R.M. (1984) An overview: real and apparent patterns in community structure. In: *Ecological Communities: Conceptual Issues and the Evidence* (D.R. Strong, D. Simberloff, C.G. Abele and A.B. Thistle, eds), pp. 3–18. Princeton University Press, Princeton, NJ.

May, R.M and Oster, G.F. (1976) Bifurcations and dynamic complexity in simple ecological models. *American Naturalist* 110, 573–99.

May, R.M. and Watts, C.H. (1992) The dynamics of predator-prey and resource-harvester systems. In: *Natural Enemies* (M.J. Crawley, ed.), pp. 431–57. Blackwell Science, Oxford.

May, R.M., Conway, G.R., Hassell, M.P. and Southwood, T.R.E. (1974) Time delays, density dependence and single-species oscillations. *Journal of Animal Ecology* 43, 747–70.

Maynard Smith, J. (1968) *Mathematical Ideas in Biology*. Cambridge University Press, Cambridge.

Maynard Smith, J. (1974) *Models in Ecology*. Cambridge University Press, Cambridge.

McCauley, E., Wilson, W.G. and DeRoos, A.M. (1993) Dynamics of age-structured and spatially structured predator-prey interactions – individual-based models and population-level formulations. *American Naturalist* 142, 412–42.

Morris, R.F. (1959) Single factor analysis in population dynamics. *Ecology* 40, 580–8.

Morris, W.F. (1993) Predicting the consequences of plant spacing and biased movement for pollen dispersal by honey bees. *Ecology* 74, 493–500.

Nee, S. (2006) Birth-death models in macroevolution. *Annual Review of Ecology and Systematics* 37, 1–17.

Nee, S., Mooers, A.O. and Harvey, P.H. (1992) Tempo and mode of evolution revealed from molecular phylogenies. *Proceedings of the National Academy of Sciences USA* 89, 8322–6.

Nicholson, A.J. (1954) An outline of the dynamics of natural populations. *Australian Journal of Zoology* 2, 9–65.

Nicholson, A.J. and Bailey, V.A. (1935) The balance of animal populations. *Proceedings of the Zoological Society Part 1 London* 3, 551–98.

Olmsted, I. and Alvarez-Buylla, E.R. (1995) Sustainable harvesting of tropical trees: demography and matrix models of two palm species in Mexico. *Ecological Applications* 5, 484–500.

Pacala, S.W. and Silander, J.A.J. (1990) Tests of neighbourhood population dynamic models in field communities of two annual weed species. *Ecological Monographs* 60, 113–34.

Park, T., Leslie, P.H. and Mertz, D.B. (1964) Genetic strains and competition in populations of *Tribolium*. *Physiological Zoology* 37, 97–162.

Pearl, R. and Reed, L.J. (1920) On the rate of growth of the population in the United States since 1790 and its mathematical representation. *Proceedings of the National Academy of Sciences USA* 6, 275–88.

Pedraza-Garcia, M. and Cubillos, L.A. (2008) Population dynamics of two small pelagic fish in the central-south area off Chile: delayed density dependence and biological interaction. *Environmental Biology of Fishes* 82, 111–22.

Phillimore, A.B. and Price, T.D. (2008) Density-dependent cladogenesis in birds. *PLoS Biology* 6, 483–9.

Pimm, S.L. (1982) *Food Webs*. Chapman and Hall, London.

Pimm, S.L. (1984) The complexity and stability of ecosystems. *Nature* 307, 321–6.

Pimm, S.L. and Lawton, J.H. (1977) Number of trophic levels in ecological communities. *Nature* 268, 329–31.

Pimm, S.L., Jones, H.L. and Diamond, J. (1988) On the risk of extinction. *American Naturalist* 132, 757–85.

Pontin, A.J. (1982) *Competition and Coexistence of Species*. Pitman, London.

Prout, T. and McChesney, F. (1985) Competition among immatures affects their adult fertility:population dynamics. *American Naturalist* 126, 521–58.

Pueyo, Y. and Beguería, S. (2007) Modelling the rate of secondary succession after farmland abandonment in a Mediterranean mountain area. *Landscape and Urban Planning* 83, 245–54.

Raghu, S. and Walton, C. (2007) Understanding the ghost of *Cactoblastis* past: historical clarifications on a poster child of classical biological control. *Bioscience* 57, 699–705.

Raup, D.M. and Sepkoski, J.J. (1982) Mass extinctions in the marine fossil record. *Science* 215, 1501–3.

Raup, D.M. and Sepkoski, J.J. (1984) Periodicity of extinctions in the geologic past. *Proceedings of the National Academy of Sciences USA* 81, 801–5.

Ricker, W.E. (1954) Stock and recruitment. *Journal of the Fisheries Research Board of Canada* 11, 559–623.

Reeve, J.D. and Murdoch, W.W. (1985) Aggregation by parasitoids in the successful control of the California red scale: a test of theory. *Journal of Animal Ecology* 54, 797–816.

Roberts, A. (1974) The stability of a feasible random ecosystem. *Nature* 251, 607–8.

Roelants, K., Gower, D.J., Wilkinson, M., Loader, S.P., Biju, S.D., Guillaume, K. et al. (2007) Global patterns of diversification in the history of modern amphibians. *Proceedings of the National Academy of Sciences USA* 104, 887–92.

Rohde, R.A. and Muller, R.A. (2005) Cycles in fossil diversity. *Nature* 434, 208–10.

Rosenzweig, M.L. and MacArthur, R.H. (1963) Graphical representation and stability conditions of predator-prey interactions. *American Naturalist* 97, 209–23.

Santangelo, G., Bramanti, L. and Iannelli, M. (2007) Population dynamics and conservation biology of the over-exploited Mediterranean red coral. *Journal of Theoretical Biology* 244, 416–23.

Scanlan, J.C., Berman, D.M. and Grant, W.E. (2006) Population dynamics of the European rabbit (*Oryctolagus cuniculus*) in north eastern Australia: simulated responses to control. *Ecological Modelling* 196, 221–36.

Schaffer, W.M. (1985) Order and chaos in ecological systems. *Ecology* 66, 93–106.

Schmid, P.E. (1992) Community structure of larval Chironomidae (Diptera) in a back water area of the River Danube. *Freshwater Ecology* 27, 151–67.

Segel, L.A. (1984) *Modeling Dynamic Phenomena in Molecular and Cellular Biology.* Cambridge University Press, Cambridge.

Seifert, R.P. and Seifert, F.H. (1976) A community matrix analysis of *Heliconia* insect communities. *American Naturalist* 110, 461–83.

Seiffert, E.R., Simons, E.L., Clyde, W.C., Rossie, J.B., Attia, Y., Bown, T.M. et al. (2005) Basal anthropoids from Egypt and the antiquity of Africa's higher primate radiation. *Science* 310, 300–4.

Sibley, C.G. and Ahlquist, J.E. (1990) *Phylogeny and Classification of Birds.* Yale University Press, New Haven, CT.

Sibly, R.M., Barker, D., Hone, J. and Pagel, M. (2007) On the stability of populations of mammals, birds, fish and insects. *Ecology Letters* 10, 970–6.

Silvertown, J. and Smith, B. (1989) Germination and population structure of spear thistle *Cirsium vulgare* in relation to experimentally controlled sheep grazing. *Oecologia* 81, 369–73.

Silvertown, J., Holtier, S., Johnson, J. and Dale, P. (1992) Cellular automaton models of interspecific competition for space- the effect of pattern on process. *Journal of Ecology* 80, 527–34.

Spencer, M. and Tanner, J.E. (2008) Lotka-Volterra competition models for sessile organisms. *Ecology* 89, 1134–43.

Steele, J. (1974) Stability of plankton ecosystems. In *Ecological Stability* (M.B. Usher and M.H. Williamson, eds), pp. 179–91. Chapman and Hall, London.

Straw, N.A. (1991) *Report on Forest Research*, p. 43. HMSO, London.

Tregonning, K. and Roberts, A. (1979) Complex systems which evolve towards homeostasis. *Nature* 281, 563–4.

Turchin, P. (1990) Rarity of density dependence or population regulation with lags? *Nature* 344, 660–3.

Turchin, P. (2003) *Complex Population Dynamics.* Monographs in Population Biology 35. Princeton University Press, Princeton, NJ.

Turchin, P. and Taylor, A.D. (1992) Complex dynamics in ecological time series. *Ecology* 73, 289–305.

Varley, G.C. (1947) The natural control of population balance in the knapweed gallfly. *Journal of Animal Ecology* 16, 139–87.

Varley, G.C. and Gradwell, G.R. (1960) Key factors in population studies. *Journal of Animal Ecology* 29, 399–401.

Verhulst, P.F. (1838) Notice sur la loi que la population suit dans son accroissement. *Correspondances Mathematiques et Physiques* 10, 113–21 [in French].

Volterra, V. (1926) Fluctuations in the abundance of a species considered mathematically. *Nature* 118, 558–60.

Volterra, V. (1928) Variations and fluctuations of the numbers of individuals in animal species living together. *Journal du Conseil Internatioanal pour l'Exploration de la Mer* III, 3–51. Reprinted in Chapman, R.N. (1931) *Animal Ecology*, appendix, pp. 409–48, McGraw-Hill, New York.

Watkinson, A.R. (1980) Density dependence in single species populations of plants. *Journal of Theoretical Biology* 83, 345–57.

Williamson, M., Gaston, K.J. and Lonsdale, W.M. (2001) The species-area relationship does not have an asymptote! *Journal of Biogeography* 28, 827–30.

Wilson, J.B. and Roxburgh, S.H. (1992) Application of community matrix theory to plant competition data. *Oikos* 65, 343–8.

Woiwod, I.P. and Hanski, I. (1992) Patterns of density dependence in moths and aphids. *Journal of Animal Ecology* 61, 619–30.

Wolfram, S. (1984) Universality and complexity in cellular automata. *Physica D* 10, 1–35.

Yule, G.U. (1924) A mathematical theory of evolution, based on the conclusions of Dr. J. R. Willis. Philosphical Transactions of the Royal Society of London Series B 213, 21–83.

Index